T0233742

Materializing the Web of Linked Data

Nikolaos Konstantinou
Dimitrios-Emmanuel Spanos

Materializing the Web of Linked Data

 Springer

Nikolaos Konstantinou
National Technical University of Athens
Athens, Greece

Dimitrios-Emmanuel Spanos
National Technical University of Athens
Athens, Greece

ISBN 978-3-319-35745-4   ISBN 978-3-319-16074-0   (eBook)
DOI 10.1007/978-3-319-16074-0

Springer Cham Heidelberg New York Dordrecht London
© Springer International Publishing Switzerland 2015
Softcover reprint of the hardcover 1st edition 2015

Printed on acid-free paper

Springer International Publishing AG Switzerland is part of Springer Science+Business Media (www.springer.com)

Preface

The Linked Open Data (LOD) paradigm is constantly gaining worldwide acceptance, due to numerous benefits yielded by its adoption. However, in order to effectively offer solutions that rely on LOD, several factors need to be taken into account, for the result to be correctly implemented and viable. In this book, we introduce the fundamental concepts stemming from the Semantic Web domain, the technological building blocks and their respective roles, and we present and analyze the various technical challenges that are associated with this effort.

The purpose of writing this book is to provide a structure when taking into account the several subproblems that have to be addressed when designing and implementing an LOD repository in a manner that allows it to play along with current systems and respective technologies. The aim is to offer a spherical view on the available implementation options regarding every logical layer and technological choice on the server side of a LOD repository. We discuss in detail the most important decisions that have to be made, regarding the methodology and choice of technologies, along with usage examples, domain-specific scenarios and the respective choice justifications.

The material is divided as follows: Chapter 1 introduces the basic concepts that are being discussed and analyzed throughout this work, Chap. 2 provides the basic technical building blocks for materializing LOD, in Chap. 3 it is shown how to put the building blocks together in order to offer more complete functionality, Chap. 4 focuses on creating Linked Data from relational databases, while Chap. 5 on creating Linked Data in real-time from sensor data streams. Both Chaps. 4 and 5 demonstrate end-to-end complete LOD solution approaches as a proof of concept. Chapter 6 concludes the book with the authors' summary and outlook.

This book is addressed to people with an interest in publishing Linked Open Data; therefore, the intended audience includes academics wishing to understand the background and rationale behind LOD publishing, content owners, data architects, IT professionals, and decision makers looking into ways to increase the value of their contents while grasping the big picture in which their dataset fits in, software engineers and Web enthusiasts in general, eager to get acquainted with state-of-the-art technologies.

Most of the hereby presented material has been presented in published research, accomplished with our colleagues, in international scientific peer-reviewed conference proceedings and journal articles. It is, though, updated and enriched, in order to bring into a uniform state several years' worth of active research in the domain.

Athens, Greece   Nikolaos Konstantinou
                                                              Dimitrios-Emmanuel Spanos

Acknowledgements

We would like to thank Iraklis Varlamis (Harokopio University of Athens) and Charalampos Doulaverakis (Centre for Research and Technology Hellas), for their time and the considerable help they offered by providing detailed comments and suggestions on the manuscript.

Contents

1	**Introduction: Linked Data and the Semantic Web**	1
	1.1 The Origin of the Semantic Web	1
	1.1.1 Why a Semantic Web?	1
	1.1.2 The Need for Adding Semantics	2
	1.2 Preliminaries	3
	1.2.1 Data, Information, Knowledge	3
	1.2.2 Interoperability, Integration, Merging and Mapping	4
	1.2.3 Semantic Annotation	7
	1.2.4 Metadata	7
	1.2.5 Ontologies	9
	1.2.6 Reasoners	10
	1.2.7 Knowledge Bases	11
	1.2.8 Linked (Open) Data	13
	References	15
2	**Technical Background**	17
	2.1 Introduction	17
	2.2 The Underlying Technologies	18
	2.3 Modeling Data Using RDF Graphs	19
	2.3.1 Using Namespaces	20
	2.3.2 RDF Serialization	21
	2.3.3 The RDF Schema	26
	2.4 Ontologies Based on Description Logics	30
	2.4.1 Description Logics	30
	2.4.2 The Web Ontology Language (OWL)	33
	2.5 Querying the Semantic Web with SPARQL	35
	2.6 Mapping Relational Data to RDF	39
	2.7 Other Technologies	44
	2.8 Ontologies and Datasets	46
	References	49

**3 Deploying Linked Open Data: Methodologies
and Software Tools** 51
   3.1   Introduction 51
   3.2   The O in LOD: Open Data 52
        3.2.1   Opening Data: Bulk Access vs. API 55
        3.2.2   The 5-Star Deployment Scheme 55
   3.3   The D in LOD: Modeling Content 56
        3.3.1   Assigning URIs to Entities 57
   3.4   Software for Working with Linked Data 61
        3.4.1   Ontology Authoring Environments 61
        3.4.2   Platforms and Environments for Working
              with (RDF) Data 63
        3.4.3   Software Libraries for Working with RDF 68
   References 70

4 Creating Linked Data from Relational Databases 73
   4.1   Introduction 73
   4.2   Motivation-Benefits 74
   4.3   A Classification of Mapping Approaches 77
   4.4   Creating Ontology and Triples
        from a Relational Database 84
        4.4.1   Creating and Populating a Domain Ontology 86
        4.4.2   Mapping a database to an existing ontology 89
   4.5   Complete Example: Linked Data from the Scholarly/Cultural
        Heritage Domain 91
        4.5.1   Synchronous vs. Asynchronous Exports as LOD
              in Digital Repositories 94
        4.5.2   From DSpace to Europeana: A Use Case 94
   4.6   Future Outlook 98
   References 99

**5 Generating Linked Data in Real-time from Sensor
Data Streams** 103
   5.1   Introduction: Problem Framework 103
   5.2   Context-Awareness, Internet of Things, and Linked Data 104
   5.3   Fusion 105
        5.3.1   JDL Fusion Levels 106
   5.4   The Data Layer 107
        5.4.1   Modeling Context 108
        5.4.2   Annotation of Sensor Data 110
        5.4.3   Real-time vs. Near-real-time,
              Synchronous vs. Asynchronous 111
        5.4.4   Data Synchronization and Timestamping 112
        5.4.5   Windowing 112
        5.4.6   The (Distributed) Data Storage Layer 113

5.5   Rule-Based Stream Reasoning in Sensor Environments 115
      5.5.1   Rule-Based Reasoning in Jena 117
      5.5.2   Rule-Based Reasoning in Virtuoso 117
5.6   Complete Example: Linked Data from a Multi-Sensor
      Fusion System Based on GSN 118
      5.6.1   The GSN Middleware 118
      5.6.2   Low Level Fusion 119
      5.6.3   A Sensor Fusion Architecture 121
      5.6.4   High Level Fusion Example 123
References 124

6 **Conclusions: Summary and Outlook** 127
6.1   Introduction 127
6.2   Discussion 129
6.3   Domain-Specific Benefits 131
6.4   Open Research Challenges 132
References 133

Abbreviations

API	Application Programming Interface
ASCII	American Standard Code for Information Interchange
CMS	Content Management System
CSV	Comma Separated Values
DAML	DARPA Agent Markup Language
DC	Dublin Core
ETL	Extract, Transform, Load
FGDC	Federal Geographic Data Committee
FOAF	Friend Of A Friend
GAV	Global-As-View
GLAV	Global–local-As-View
GRDDL	Gleaning Resource Descriptions from Dialects of Languages
GSN	Global Sensor Networks
HLF	High Level Fusion
HTML	HyperText Markup Language
HTTP	HyperText Transfer Protocol
IoT	Internet of Things
IRI	Internationalized Resource Identifier
JDL	Joint Directors of Laboratories
JSON	JavaScript Object Notation
LAV	Local-As-View
LDP	Linked Data Platform
LLF	Low Level Fusion
LOD	Linked Open Data
MPEG	Moving Picture Experts Group
OAI-PMH	Open Archives Initiative Protocol for Metadata Harvesting
OBDA	Ontology-Based Data Access
OIL	(Ontology Inference Layer) or (Ontology Interchange Language)
OMG	Object Management Group
OWL	Web Ontology Language

PHP	Hypertext Preprocessor (recursive backronym). Originally stood for Personal Home Page
POWDER	Protocol for Web Description Resources
R2RML	RDB to RDF Mapping Language
RDF	Resource Description Framework
RDFS	RDF Schema
REST	REpresentational State Transfer
RFID	Radio-Frequency IDentification
RIF	Rule Interchange Format
RSS	Really Simple Syndication (RSS 2.0) or RDF Site Summary (RSS 1.0 and RSS 0.90) or Rich Site Summary (RSS 0.91)
RTP	Real-time Transport Protocol
SIOC	Semantically-Interlinked Online Communities
SKOS	Simple Knowledge Organization System
SOAP	Simple Object Access Protocol
SPARQL	SPARQL Protocol and RDF Query Language (recursive acronym)
SPIN	SPARQL Inferencing Notation
SQL	Structured Query Language
STO	Situation Theory Ontology
SVG	Scalable Vector Graphics
TSV	Tab Separated Values
UML	Unified Modeling Language
URL	Uniform Resource Locator
URI	Uniform Resource Identifier
VoID	Vocabulary of Interlinked Datasets
W3C	World Wide Web Consortium
WSDL	Web Services Description Language
XHTML	eXtensible HyperText Markup Language
XML	eXtensible Markup Language
XSD	XML Schema Definition
XSL	eXtensible Stylesheet Language

Chapter 1
Introduction: Linked Data and the Semantic Web

1.1 The Origin of the Semantic Web

In order to understand the Linked Data concept, it is important to have first a brief discussion about the Semantic Web vision. We will see next that the Semantic Web concept and the Linked Data concept are closely interrelated, as the latter implements the vision of the former. The term "Semantic Web" was primarily coined by the inventor of the Web itself, Tim Berners-Lee et al. (2001). The main idea is that Web content, in order to be understandable both by humans and software, it should incorporate its meaning, its *semantics*.

This chapter provides some historical background and justification for the creation of the Semantic Web and Linked Data concepts (Sect. 1.1) and an introduction to the most important concepts (Sect. 1.2) which we discuss throughout this work.

1.1.1 Why a Semantic Web?

Among the most important problems that were raised with the rapid evolution of the Internet, included dealing with information management: Sharing, accessing, retrieving it and most importantly consuming it becomes a task increasingly tedious. Evolution in the search engines domain allowed web users access to exabytes of data in billions of web pages, with information for almost anything; pages that would be impossible to discover and search without the use of special Information Retrieval (IR) techniques offered by current search engine implementations.

Nevertheless, simple keyword-based searching is a solution that does not release the potential in the information available in the Web. This is because firstly, there is a large quantity of existing data on the web not stored in HTML but in other technologies, such as relational databases. This is content that cannot be discovered and crawled by search engines. This information, often referred to as the Deep Web (Bergman

© Springer International Publishing Switzerland 2015
N. Konstantinou, D.-E. Spanos, *Materializing the Web of Linked Data*,
DOI 10.1007/978-3-319-16074-0_1

2001) does not come online until it is generated dynamically in response to a direct request. As a consequence, traditional search engines cannot retrieve this content.

Secondly, and most importantly, web pages do not carry any meaning regarding their contents. Therefore, conventional searching relies only on keyword matching and does not take into account the meaning of the content.

In order to overcome these obstacles in releasing the information potential, the creator of the Web, Tim Berners-Lee, proposed the idea of the Semantic Web. The Semantic Web is about bringing "structure to the meaningful content of Web pages, creating an environment where software agents, roaming from page to page, can readily carry out sophisticated tasks for users". The idea is not to replace the current web, but rather to complement it by constituting an addition to the existing content in a way that will allow integration of this content into an exploitable source of knowledge. The initial article by Tim Berners-Lee in Scientific American was describing an anticipated evolution of the existing Web into a Semantic Web (Berners-Lee et al. 2001). Since then, however, an emerging web of interconnected published datasets in the form of Linked Data implements the Semantic Web vision, the Web of Data (Bizer et al. 2009).

1.1.2 The Need for Adding Semantics

First, let us introduce the meaning of *semantics*: The term is typically used to denote the study of meaning. The word comes from the Greek work "σημαντικός" (pronounced "simantikós") that stands for "significant". Usage of semantics means to try to capture the interpretation of a formal or natural language. This enables entailment, application of logical consequence. This logical consequence is about relationships between the statements that are expressed in this language. In Chap. 2 we will have the chance to delve deeper into formal languages and expressions of concepts.

Now, the notion of *syntax* is a completely different concept, not to be confused with semantics. Contrarily to the notion of semantics, the notion of syntax is the study of the principles and processes by which sentences can be formed in languages. Interestingly, the word syntax is also of Greek origin ("συν" and "τάξις", pronounced "sin" and "táxis", meaning "together" and "ordering", respectively).

It is therefore possible that in a language we can have two statements that are syntactically different but semantically equivalent, in the sense that the meaning that is deduced by them is the same. Simply put, there can be more than one ways to state an assumption.

Now, why are all these related to information on the Web? As mentioned before, traditional search engines rely on keyword matches. Of course, keyword-based technologies have come a long way since the dawn of the internet, but more complex queries, taking into account the semantics of concepts being searched, are still partially covered. While keyword-based queries will return accurate results, the vast variety of the information that will be available on the internet will not even be queried. Searching for specific information using methods that do not include or support semantics is not guaranteed to work as expected: Query hits are based on keyword matches, a result of extraction and indexing of keywords contained in web pages, in order to accommodate user searches. This usually leads to low precision in fetching the results. In addition, important results may not be fetched, or ranked low, as there may not be exact keyword matches. As a result, keyword-based search engines do not return usually the desired results, despite the vast amounts of information they may retrieve. This is demonstrated by the typical user behavior: often the users prefer to change their query than navigate beyond the first page of the search results.

Let us suppose for an instant that we are fine with that: a researcher, an interested visitor would at least have a list of references, web pages to navigate to in order to collect more information. This exactly is the second aspect of the problem, as the search will return web pages that will have to be processed by a human. They are, therefore, suitable for human consumption and not for machine consumption. This is because of the lack of structure and semantics from their contents, despite the links that may be existing, from one web page to another.

The Semantic Web—among others—aims at efficiently tackling these issues, and achieves this by offering ways to describe of Web resources. In practice, the goal is to enrich existing information by the addition of semantics that specify the resources in a way understandable both by humans as well as by the computers. This will make information easier to discover, classify and sort, with comparison to approaches that do not incorporate semantics. This semantic enrichment enables semantic information integration and subsequently increases serendipitous discovery of information. Moreover, by having machine-understandable information, automated software agents are able to consume and understand the information in a manner that allows inference.

Making use of semantics, the researcher of the example would not have to visit every result page separately, distinguishing the relevant from the irrelevant results. Instead, he or she would retrieve a list of more related information, in a machine-processable manner, ideally containing links to more relevant information.

1.2 Preliminaries

In order to understand how the previously described vision turns into a reality, several concepts need first to be introduced.

1.2.1 Data, Information, Knowledge

The terms *data* and *information* are concepts that are often used interchangeably. According to the Cambridge English Dictionary, data is: "Information, especially facts or numbers, collected to be examined and considered and used to help decision-making, or information in an electronic form that can be stored and used by a computer", while information concerns "facts about a situation, person, event, etc.". In Computer Science, the smallest information particle is the *bit*, which replies with a yes or no (1 or 0). A bit can carry data, but in the same time, it can carry the result of a process, for instance whether an experiment had a successful conclusion or not.

Since these terms are fundamental to this book, it needs to be clarified that their meaning, as close as it may be, is slightly different. This small difference is purely subjective and depends on our viewpoint. Data is collected in order "to be examined and considered and used", while information is considered to contain facts. We could state that the information is the outcome of data processing. Information is the meaning that is assigned to a data set which has been processed using predefined rules having produced several useful conclusions.

Therefore, something that is considered as "data" for one observer, can be considered as "information" by another, according to their point of view. As an example, consider the value of a field, e.g. "temperature" in a database that collects sensor measurements. This value could be the information produced after statistical analysis over numerous measurements by a researcher wishing to extract the average value in a region over the years. For another researcher, this could be a part of the data that will contribute in drawing conclusions regarding climate change.

Similarly, *knowledge* is defined by the Cambridge English Dictionary as "understanding of or information about a subject that you get by experience or study, either known by one person or by people generally". The key term here is "by experience or study" which, again, implies processing of the underlying information.

As a conclusion, we cannot draw clear lines between data, information and knowledge. For this reason, in this book, the term that is used each time depends on the respective point of view.

We could say that when we process collected data in order to extract meaning from it, we create new information based on it. The addition of semantics is indispensable in generating new knowledge based on this information. This process is the semantic enrichment of the information and makes it unambiguously understood by any interested party (by people generally).

1.2.2 Interoperability, Integration, Merging and Mapping

Linked Data is (among others) about linking the data, typically in order to process it. This data that may reside in distributed, interconnected or not, data sources. Therefore, in order to manage and process it, it is often required to enable interoperability among the systems that store it.

A key concept here is the data *schema*. A data schema defines how the data is to be structured. The word comes from the Greek word σχήμα (pronounced "schíma"), which means the *shape*, or, more generally, the *outline*. The schema can be regarded as a common agreement regarding the interchanged data.

Now, two systems can be considered as *interoperable* when they are in position to successfully exchange information between them (Uschold and Gruninger 2002). According to (Park and Ram 2004), existing approaches that achieve interoperability among distributed software clients can be broadly categorized as follows:

- *Mapping* among the concepts of each source. A global information description schema translates the concepts from one source to another.
- *Intermediation* in order to translate queries. According to this approach, an intermediate layer is inserted, in order to translate the queries according to each data source's schema. This layer can be an additional piece of software, a set of rules, an ontology, a software agent, etc.
- *Query-based* approach. In this approach, the user is able to pose queries that will be evaluated on every data source.

These approaches are not mutually exclusive. For instance, a system can include intermediators while also having a global schema to describe information.

In all cases, protocols and standards play a crucial role in enabling interoperability. Attempts such as SOAP and WSDL as long as a variety of microformats from W3C, OMG and other consortia facilitate communication and therefore interoperability. Adoption by all involved data sources of a common standard to represent information is very important in interoperability.

It has to be mentioned that in most cases, because of the openness of the Web it is impossible to impose use of standards. In these cases, W3C *recommendations* are as close as possible to standards, in the sense that the term has in other domains.

Information *integration* is an approach fundamentally different than interoperability. Integration is a term often confused with interoperability, while the latter only refers to the successful information exchange between two systems. In order to achieve integration, simple compliance to standards in order to enable communication is not sufficient. Integration, in the context of Computer Science is the process according to which the information that originates from various sources and systems, is combined in order to allow its processing as a whole.

In this book, *information integration* refers to processing and handling as a whole information that stems from heterogeneous sources of storage and processing. This heterogeneity is a result of the numerous ways in which nowadays information can be stored, organized and communicated. In particular, the heterogeneity problem can comprise several of the following problems:

- Different models and schemas. For instance, information may reside in a relational database, XML files or a Knowledge Base.
- Differences in the vocabulary. For instance, the property "time" in one vocabulary may appear as "duration" in another.
- Syntactic mismatch. The same piece of information may appear at the path `/appearance/description` in one XML file and in path `/appearance/@description` or even `/description/appearance` in another.
- Semantic mismatch. For instance, in the RDF model the concept of superclass/subclass exists while in the relational mode it does not.

In general, existing information integration approaches can be broadly categorized in two major categories: According to the first approach, every element of the global schema is expressed as a query/view over the schemas of the sources. This approach is called *Global-As-View*. This approach is preferable when the source schemas are not subject to frequent changes. According to the second approach, every element of the local schemas is expressed as a query/view over the global schema (*Local-As-View*). The approach in which there are mappings among the sources but a common schema is absent is referred to as P2P. Ideally, the architecture of an information integration system is as follows:

A source Π_1 with schema S_1, a source Π_2 with a schema S_2, ..., a source Π_ν with source S_ν, and a global schema S in which the higher level queries are posed. The ultimate goal in the information integration problem is the provision to the user of the ability to submit queries to the schema S and receive answers from $S_1, \ldots S_\nu$, without having to deal with or even being aware of the heterogeneity in the information sources. The benefit that is earned is that the global schema S is in position to answer higher level queries, contrarily to the source schemas. In any case, what is desired is the integrated access to the information, regardless to where it is stored and who administers it.

Now, *semantic information integration* is the addition of its semantics in the resulting integration scheme.

It has to be noted that the term information *merging* is a term different than information integration as the former, contrarily to the latter, implies unifying the information at the implementation/storage layer.

Contrarily, the concept of *mapping* is tightly coupled with the concept of integration. In data mapping, a mapping is the specification of a mechanism through which the members of a model are transformed to members of another model, which conforms to a meta-model that can be the same, or different. A mapping can be declared as a set of relationships, constraints, rules, templates or parameters that are defined during the mapping process, or through other forms that have not yet been defined. Now, a data integration system can be defined as a triple:

$$I = \langle G;S;M \rangle$$

where

- G is the global schema,
- S is the source schema and
- M is a set of mappings between G and S.

In the Local-As-View approach, each declaration in M maps an element from the source schema S to a query (a view) over the global schema G. In the Global-As-View, the opposite takes place: every declaration in M maps an element of the global schema G to a query (a view) over the source schema S. The global schema G is a unified view over the heterogeneous set of data sources. In general, the goal is always the same: unifying the data sources under the same common schema.

Using the definition of data integration systems, we can define the concept of mapping: A mapping m (member of M) from a schema S to a schema T is a declaration of the form $Q^S \rightsquigarrow Q^T$ where Q^S is a query over S and Q^T a query over T (Lenzerini 2002).

1.2.3 Semantic Annotation

The term *annotation* essentially is about adding metadata to the data. The data itself can be encoded in any standard. The value of the annotation is especially important in cases when data is not human-understandable in its primary form, as it happens e.g. with multimedia.

The specification *semantic* expresses the addition of the semantics in the annotation (i.e. the metadata), in a way that the annotation will be machine-processable in order to infer additional knowledge. The difference between semantic annotation and simple annotation lies in the fact that the former describes the data using a common, established way while the latter using keywords or other ad hoc serialization which impedes further processing of the metadata.

Given the importance of annotation, one could wonder why it is not always present. The answer is a combination of several factors: First of all, it is a time-consuming process. Users simply do not have enough time or do not consider it important enough in order to invest time to annotate their content. Moreover, in order to semantically annotate content the user performing the annotation must be at least familiar with both the technical as well as the conceptual part regarding the annotation. There is also the risk of outdated annotations, especially in rapidly changing environments with lack of automation in the annotation process.

Automation in annotating content is in general a step to the right direction, especially since incomplete annotation is preferable to absent. However, systems performing automated annotation (e.g. in video streams) often present limited recall and precision (lost or inaccurate annotations).

1.2.4 Metadata

Metadata is essentially data about data. This can be materialized efficiently with the use of ontologies, as we will see next, since ontologies can be used to describe Web resources by adding descriptions about their semantics and their relations.

Describing Web resources using ontologies aims among others to make them machine-understandable (Berners-Lee et al. 2001). In order to achieve this, it is necessary that each (semantic) annotation be corresponded to a piece of information. It is important in this process that the annotation of Web resources follows a common standard.

This annotation is commonly referred to as "metadata". Usually, metadata are provided in a semi-structured form, in the sense that they are provided in separate files that contain the metadata themselves as well as their structure. Semi-structured data sources are sources in which the structure accompanies the metadata, in contrast to structured sources, in which the structured is stored separately as is the case with e.g. relational databases.

Popular languages to describe semi-structured data include XML and JSON, which serve efficiently metadata description and presentation needs. These languages offer a strictly defined syntax and they can be easily customized to suit the needs of virtually any application. Related libraries exist in almost any programming language that make XML or JSON processing a typical, automated task.

Several software solutions have been developed that utilize XML capabilities. For instance, we find XML files in configuration files, information that is sent via Web Services in the form of SOAP envelopes, content that is made available in RSS form, even as the base for generating content in various formats, using XSL.

Among the powerful characteristics of these languages is the fact that the documents in these languages are human-understandable and machine-processable. The practically unlimited hierarchical structure covers most of the description requirements that may occur. Moreover, storing it in separate files allows its collaboration with communication protocols, which makes XML files even more manageable. Finally, the files are independent from the environment in which they reside, a property that makes them resilient to technological evolutions.

However, this way of describing metadata presents several negative points. Among these, is the fact that the description model is not expressive enough, at least not as expressive as the model in relational databases. Also, the model has a limited way of structuring the information.

On the other hand, ontology description languages offer a richer way of describing information, both on the web and offline. For instance, using ontology descriptions, we can define relationships among concepts such as subclass/superclass, mutually disjoint or inverse concepts etc. (to be analyzed next in more detail). In these cases, the role of having a semi-structured representation of these metadata is restricted into merely providing a uniform syntax for these metadata files (Decker et al. 2000).

When the nature of the data requires inclusion of semantics, current metadata need to be replaced/updated with newer ones, with more expressive capabilities and terminology. By looking at the Web landscape more closely, the introduction and adoption of the RDF standard was driven by the need for more descriptive standards than XML. The Resource Definition Framework, can be regarded as the evolution of XML, which adds powerful capabilities to this direction. RDF was mainly designed to describe Web Resources, however its capabilities extend far beyond that. It is possible to express RDF using XML or JSON syntax, or other syntaxes, as explained in Sect. 2.3.2.

Targeting at a more comprehensive, precise and consistent description of Web Resources, the Web Ontology Language (OWL) has been developed (see Sect. 2.4.2).

1.2.5 Ontologies

Simply put, an ontology is an additional description to an unclear model that aims to further clarify it. Ontologies provide conceptual description models that can be understood both by human and computer, under the form of a concept description. An ontology is responsible for the description of a specific domain, by describing its related concepts and their relationships. Their use aims at bridging and integrating multiple and heterogeneous digital content on a semantic level, which is exactly the key idea of the Semantic Web vision. Ontologies provide conceptual domain models, which are understandable to both human beings and machines as a shared conceptualization of a given specific domain.

The term has been borrowed by Philosophy, according to which an "Ontology" is a systematic recording of "Existence".[1] For a software system, something that "exists" is something that can be represented which means in general that the set of concepts that can be represented is the universe of discourse that is described by the ontology.

A definition of what is an ontology has been given by Gruber (1995) as "the specification of a conceptualization". A *conceptualization* is defined as a *structure* D,R where D is a *domain* and R a set of relevant relations on D (Genesereth and Nilsson 1987). The set of relations comprises the *intensional* and the *extensional* relations also referred to as *conceptual* and *ordinary* relations, respectively. The *domain space* is defined as a *structure* D,W, where W is a *set* of maximal states of affairs of such domain (also called *possible worlds*). A *conceptualization* for D can then be defined as an ordered triple set $C = D,W,R$, where R is a set of conceptual relations on the domain space D,W. More simply, an ontology can be described as a set of definitions that associate the names of entities in the universe of discourse with human-readable text describing the meaning of the names and a set of formal axioms that constrain the interpretation and well-formed use of these terms.

According to the degree of generalization, we can distinguish several types of ontologies:

1. A *top-level* ontology defines very abstract, general terms such as time, space, property, event etc. which are valid regardless of the domain of interest.
2. A *domain* ontology defines concepts that are related to a specific knowledge domain, such as the medical domain, or even a narrower domain such as e.g. digital cameras.
3. A *task* ontology defines concepts that are related to accomplishing an activity, such as assembling an item using several parts.
4. An *application* ontology is restricted to a specific application and defines only the concepts needed for it.

---

[1] It is generally accepted that the term "Ontology", with a capital "O" refers to the Philosophy branch, while the term "ontology", with a small "o" refers to Semantic Web ontologies in Computer Science.

Of course, clear lines among these categories cannot be drawn, neither is there any formal specification. The idea is to illustrate that ontologies can be used to model concepts regardless to how general or specific these concepts are.

With the use of ontologies, content is made suitable for machine consumption, contrary to the majority of the content found on the Web today, which is primarily intended for human consumption. The use of an ontology in content description can lead to an automated increase of the system knowledge, by using logical rules that can be applied to the existing knowledge in order to infer facts that are implicitly declared (Baader et al. 2003) (see Sect. 2.6).

1.2.6   Reasoners

A *reasoner* (or reasoning engine) is a software component whose primary goal is to validate the consistency of an ontology, by performing consistency checks on it and also, to infer knowledge that is implicitly declared. This is typically done by applying simple rules of deductive reasoning. Therefore, a reasoner in practice can augment a Knowledge Base by extracting the implicitly defined information (inferring facts that have not been explicitly stated), according to the rules that have been defined.

Reasoning is the procedure of extracting knowledge that has not been explicitly stated in the system. It is also among the most powerful features of semantic web technologies: reasoning upon the knowledge *explicitly* stated allows inference of knowledge that is *implicitly* stated.

Furthermore, a reasoner inspects an ontology model and can check its consistency, satisfiability and classification of its concepts (Donini et al. 1996), leading to the existence of a "healthy" and robust context model that does not contain any disaccords among its term definitions.

It is not possible to produce inferred knowledge without applying reasoning procedures. The basic reasoning procedures can be categorized as follows (Donini et al. 1996):

- *Consistency checking.* Assures that the ontology does not contain contradictory facts. According to the DL terminology, this service assures that the set of facts that refer to specific individuals (Assertional Box or ABox) will be consistent with respect to the set of axioms that describe the general concepts of a domain (Terminological Box or TBox).
- *Concept satisfiablility.* Checks whether a class can have individuals. If a class, because of its definition, is not in position to contain individuals, this means that the entire ontology is not consistent.
- *Concept subsumption*, classification. It can calculate the hierarchy tree of classes/ subclasses. This tree can be used in order to evaluate queries over a class and return inductively its superclasses or subclasses.

- *Instance checking (Realization)*. It can calculate the classes to which an individual belongs. In other words, it can calculate the individual type. Realization can take place only after classification, as the concept hierarchy is necessary.

Regarding reasoning properties now, as far as Description Logics (and Logics in general) are concerned, three properties are of special interest.

1. *Termination*. Declares whether the algorithm can terminate or there is the possibility of being executed indefinitely.
2. *Soundness*. A reasoning procedure is sound, if for every formula that proves to be satisfiable, then the formula is indeed satisfiable.
3. *Completeness*. A reasoning procedure is complete, if it is able to deduce every possible fact that can be inferred from the available set of axioms.

Reasoning is an important and active field of research, investigating the tradeoff between the expressiveness of ontology definition languages and the computational complexity of the reasoning procedure, as well as the discovery of efficient reasoning algorithms applicable to practical situations. This is the main rationale behind suggesting various "flavors" of the OWL language (see Sect. 2.4.2).

There is a variety of available reasoners, commercial ones like RacerPro (Haarslev et al. 2012) or free of charge like the open-source HermiT, Pellet (Sirin et al. 2007) and FaCT++ (Tsarkov and Horrocks 2006), which have different features and performance characteristics according to the application specific needs, e.g., in case of voluminous knowledge bases, a reasoner that scales well to millions of facts should be used. All these reasoning servers can function stand-alone and communicate via HTTP with the deployed system, leaving the reasoner choice up to the user.

1.2.7 Knowledge Bases

According to the Description Knowledge Handbook (Baader and Nutt 2002), a Knowledge Base (KB) is: a Graph (ABox, TBox) in combination with a reasoner engine. The main difference between an ontology and a KB is that the KB offers reasoning as an addition to the model. Reversely, an ontology can be seen as a model of a KB, without reasoning capabilities. The presence of a reasoner is indispensable to a KB, since it is the module that spawns new knowledge: the reasoner is capable of inferring facts from the facts explicitly defined in the ontology (sometimes referred to as axioms). Thus, it is not possible to extract implicit knowledge from the ontology without the use of reasoning procedures (Borgida et al. 2002).

As illustrated in Fig. 1.1, a KB that is created using Description Logics consists of two parts. The first part contains all the concept descriptions, the *intensional* knowledge and is called Terminological Box (TBox). The TBox introduces the terminology and the vocabulary of an application domain. It can be used to assign

Fig. 1.1 Elements of a
knowledge base (Baader and
Nutt 2002)

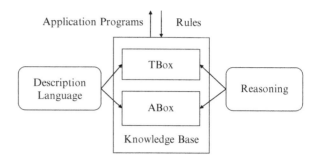

names to complex descriptions of concepts and roles. The classification of concepts is done in the TBox by determining subconcept/superconcept relationships between the named concepts. It is then possible to form the subsumption hierarchy. The second part contains the real data, the *extensional* knowledge and is called Assertional Box (ABox). The ABox contains assertions about named individuals in terms of the vocabulary defined in the TBox (Baader and Nutt 2002).

1.2.7.1   Knowledge Bases Vs. Databases

Both knowledge bases and databases are used in order to maintain models and data of some domain of discourse. Therefore, they could be considered similar. There are, however, great differences between them.

A naive approach would be to consider that the TBox of the ontology corresponds to the schema of the relational database and that the ABox corresponds to the schema instance. However, things are more complex than that. The relational model supports only untyped relationships among relations, while the ontological scheme allows for more complex relationships to be stated. In fact, the relational model does not provide enough features that could be used to assert complex relationships among data, while knowledge bases can provide answers about the model that have not been explicitly stated to it, with the use of a reasoner.

Another difference is the kind of semantics that each schema holds. The relational database schema abides by the so called "closed world assumption". The system's awareness of the world is restricted to the facts that have been explicitly stated to it. Everything that has not been stated as true fact is false. In a "closed world", a null value about a subject's property denotes the non-existence i.e. a NULL value in the isCapital field of a table Cities claims that the city is not a capital. The database answers with certainty because, according to the closed world assumption, what is not currently known to be true is false. Thus, a query like "select cities that are capitals" will not return a city with a null value at a supposed boolean isCapital field.[2] On the other hand, a query on a KB can return three types of answers: true, false and cannot tell. The open world assumption

---

[2] The example is inspired by the one found in en.wikipedia.org/wiki/Open_World_Assumption, accessed on 10-11-2014.

states that lack of knowledge does not imply falsity. In this case, information that is not explicitly declared as `true` is not necessarily `false`, it can also be `unknown`. So a question "Is Athens a capital city?" in an appropriate schema will return "cannot tell" if the schema is not informed while a database schema would clearly state `false`, in the case of a null `isCapital` value.

An interesting feature that is present in knowledge bases is the one of *monotonicity*: A system is considered monotonic when new facts do not discard existing ones. Since this feature is typically present in knowledge bases, the first version of the SPARQL query language for RDF graphs (introduced in Sect. 2.5) did not include update/delete functionality. However, in order to meet real world needs, implementations included these functions, resulting to their inclusion in SPARQL 1.1. The recommendation describes that these functions should be supported, however, it does not describe the exact behaviour of the system implementing the specification, in cases of update/delete functions.

Finally, knowledge bases and databases are also technologies of different capabilities, a fact that naturally leads to different intended use. Databases are typically used in order to manipulate large and persistent models of relatively simple data while knowledge bases typically contain fewer but more complex data.

1.2.8   Linked (Open) Data

The Semantic Web ecosystem nowadays finds application in many cases out of the academia. Technology maturity in the domain has enabled publishing and exploiting semantically annotated datasets, leading to the creation of a wider movement towards *Linked Open Data* (*LOD*), a term that has turned out into a buzzword itself, and has evolved a lot since Tim Berners Lee's initial design issues.[3]

1.2.8.1   The LOD Cloud

We are currently witnessing a proliferating number of semantically annotated repositories. For instance, in the government data domain, more and more governments provide such repositories, in an open form, in order to increase transparency, effectiveness, accuracy and citizen participation in the scope of an open and connected governance landscape. Examples of open state/government data include http://data.gov (USA), http://data.gov.uk (UK), http://data.gov.au (Australia), http:// opengov.se (Sweden) and much more, comprising a list too long to be listed here. While the aforementioned are cases of Open Data, this does not necessarily mean that they are linked.

Linked Open Data (LOD) means publishing structured data, in an open format, and making it available for everyone to use it: it is about publishing on the Web and

---

[3] Tim Berners-Lee: Linked Data Design Issues: www.w3.org/DesignIssues/LinkedData

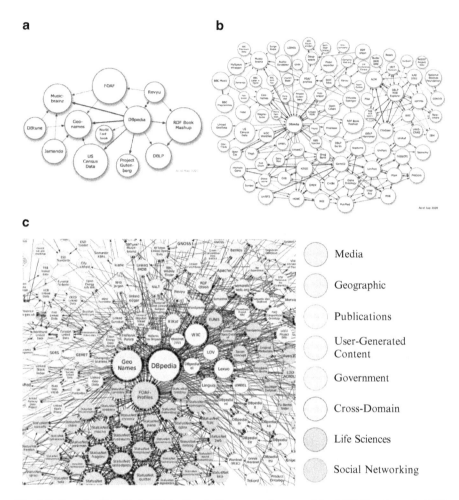

Fig. 1.2 Growth of the Linked Data Cloud. We can see the state of the LOD Cloud state in 2007, 2009, and 2014, in (**a–c**), respectively (*source*: lod-cloud.net)

connecting, using Web technologies, related data that was not previously linked, or was linked using other methods. Using URIs and RDF for this is very convenient because the data can be interlinked, creating a large pool of data, offering the ability to search, combine and exploit the knowledge. Users can even navigate between different data sources, following RDF links, and browse a potentially endless Web of connected data sources (Bizer et al. 2009).

Now, datasets that have been published while forming links among them, span several domains of human activities, forming a cloud that extends much more beyond government data, the LOD cloud. Figure 1.2 illustrates the evolution of the LOD cloud throughout the latest years. The LOD diagram serves the purpose of demonstrating the core idea and provides a visual about the core idea behind the LOD cloud generation. Overall, it is anticipated that the LOD cloud will gradually increase in size thus providing opportunities for access to data and knowledge. As depicted in Fig. 1.2, the LOD cloud is constantly increasing in terms of volume, meaning that several organizations worldwide, from various sectors, recognize the need for publishing their data as LOD and the benefits this approach has for them.

Figure 1.2c shows that the published crawlable Linked Data datasets span several domains, most notably the Media, Geographic, Publications, User-generated content, government, life sciences, social networking domains, with the addition, of course, of cross domain datasets. A collection of Semantic Web Case Studies and Use Cases can be found at: www.w3.org/2001/sw/sweo/public/UseCases/. Also, datacatalogs.org contains a list of (382 to date) catalogs of open data, curated by experts from around the world.

References

Baader F, Nutt W (2002) Basic description logics. In: Baader F, Calvanese D, McGuinness DL et al (eds) Description logic handbook. Cambridge University Press, Cambridge, pp 47–100

Baader F, Horrocks I, Sattler U (2003) Description logics as ontology languages for the Semantic Web. In: Hutter D, Stephan W (eds) Mechanizing mathematical reasoning. Lecture notes in artificial intelligence (Lecture notes in computer science), vol 2605. Springer, Heidelberg, pp 228–248

Bergman MK (2001) The deep web: surfacing hidden value. J Electron Publ 7(1). doi:10.3998/3336451.0007.104

Berners-Lee T, Hendler J, Lassila O (2001) The Semantic Web—a new form of Web content that is meaningful to computers will unleash a revolution of new possibilities. Sci Am 284(5):34–43

Bizer C, Heath T, Berners-Lee T (2009) Linked Data—the story so far. Int J Semant Web Inf Syst 5(3):1–22

Borgida A, Lenzerini M, Rosati R (2002) Description logics for databases. In: Baader F, Calvanese D, McGuinness DL et al (eds) Description logic handbook. Cambridge University Press, Cambridge, pp 472–494

Decker S, Melnik S, van Harmelen F et al (2000) The Semantic Web: the roles of XML and RDF. IEEE Internet Comput 4(5):63–74

Donini F, Lenzerini M, Nardi D et al (1996) Reasoning in description logics. In: Brewka G (ed) Principles of knowledge representation. CSLI/Cambridge University Press, Cambridge, pp 191–236

Genesereth MR, Nilsson NJ (1987) Logical foundations of artificial intelligence. Morgan Kaufmann, San Francisco

Gruber T (1995) Toward principles for the design of ontologies used for knowledge sharing. Int J Hum Comput Stud 43(5–6):907–928

Haarslev V, Hidde K, Möller R et al (2012) The RacerPro knowledge representation and reasoning system. Semant Web 3(3):267–277

Lenzerini M (2002) Data integration: a theoretical perspective. In: Proceedings of the 21st ACM SIGMOD-SIGACT-SIGART symposium on principles of database systems (PODS'02), Madison, June 2002. ACM, pp 233–246

Park J, Ram S (2004) Information systems interoperability: what lies beneath? ACM Trans Manag Inf Syst 22(4):595–632

Sirin E, Parsia B, Grau B et al (2007) Pellet: a practical OWL-DL reasoner. J Web Semant
    5(2):51–53
Tsarkov D, Horrocks I (2006) FaCT++ description logic reasoner: system description. In:
    International joint conference on automated reasoning (IJCAR 2006). Lecture notes in artifi-
    cial intelligence (Lecture notes in computer science), vol 4130. Springer, Heidelberg,
    pp 292–297
Uschold M, Gruninger M (2002) Creating semantically integrated communities on the World Wide
    Web. Invited Talk Semantic Web Workshop Co-located with WWW 2002, Honolulu

Chapter 2
Technical Background

2.1 Introduction

Linked Data is not a single technology, it is rather a set of technologies. These technologies are focused on the Web, since they use the Web as a storage and communication layer and they aim in providing meaning to the Web content. Linked Data refers to Web content since there is no value in adding semantic annotations to items that are offline, e.g. to the files on a user's desktop. That is of course, unless the intent is to publish them online at a later stage. Moreover, semantics only add value when being part of a larger context, without the presence of which, there is no need for semantics. In order to provide this context, the graph approach is employed, in which data is modeled as a graph (see Sect. 2.2).

In this chapter, we provide first the fundamental theoretical background, which serves as a basis for modeling and describing data and knowledge. Next, we present the technologies upon which the Linked Data ecosystem is built. As we will see, in the Semantic Web world, implementing a knowledge model, both resulting in graphs consisting of data triples, i.e. structured statements about resources. The respective theoretical and technical background is provided. We see how using RDF allows describing the world using a graph-based model. Next, we see how the world can be described using OWL, an approach based on totally different theoretical foundations but structurally compatible to RDF. Sections 2.3 and 2.4 offer an introduction to these concepts. The chapter continues by providing the most important technologies used in order to query (Sect. 2.5) and map RDF content to relational databases (Sect. 2.6) and some other technologies that altogether comprise the Semantic Web technology ecosystem (Sect. 2.7).

When the goal is to build a dataset that will be made part of a larger ecosystem, several conventions need to be met regarding the vocabulary, since the content has to be annotated using other people's vocabularies, something that may not feel natural to the user. However, as we see in Sect. 2.8, instead of using arbitrary fields to store information, it makes more sense to reuse widespread vocabularies and

© Springer International Publishing Switzerland 2015
N. Konstantinou, D.-E. Spanos, *Materializing the Web of Linked Data*,
DOI 10.1007/978-3-319-16074-0_2

ontologies. The chapter concludes with presenting these common vocabularies that can be reused, and adapted to new data needs, in order to maximize the data's outreach potential.

2.2 The Underlying Technologies

The foundations for Linked Data have been laid by technologies that are fundamental to the Web: HTTP (HyperText Transfer Protocol) and IRIs (URL is subset of URI, which is subset of IRI).

- HTTP is an application protocol for the management and transfer of hypermedia documents in decentralized information systems. It defines a number of request types and the expected actions that a server should carry out when receiving such requests. The HTTP protocol serves as a mechanism to serialize resources that can be serialized as a stream of bytes, such as a photo of a person, or retrieving *descriptions* about resources that cannot be sent over network, such as the person itself.
- Such resources are identified by URLs (Uniform Resource Locators), which unambiguously identify a *location* where a document about that resource can be found. URIs (Uniform Resource Identifiers) are a superset of URLs in the sense that while URLs are addresses of documents and other entities that can be found online, URIs provide a more generic means to identify anything that exists in the world. IRIs are internationalized URIs and they extend upon URIs by using the Universal Character Set whereas URIs are limited to the far fewer ASCII characters.

The documents delivered through HTTP are usually expressed in HTML (HyperText Markup Language), a markup language for the composition and presentation of various types of content (text, images, multimedia) into web pages. HTML has gone through several development phases, leading to the current HTML5 recommendation,[1] which mostly extends the support for multimedia and mathematical content through the introduction of additional markup tags.

Another foundational technology for the Web is XML (eXtensible Markup Language), which allows for strict definition of the structure of information through the use of markup tags. Several web technologies are XML-based or have a syntax or profile that is based on XML, as is the case for HTML5 for which the XHTML5 syntax is defined. A valid XHTML5 document also constitutes a valid XML document. The RDF model, which plays central role in the Semantic Web, also follows XML syntax.

In the Linked Data world, HTTPs and URIs are complemented by a technology that is essentially the cornerstone of the Semantic Web: RDF. While HTML, conveyed over HTTP, provides the means to create web documents and URIs to

---

[1] HTML5: www.w3.org/TR/html5/

create links between these documents, RDF has a more generic purpose. It provides a generic data model in which the things that are being described can be structured in order to form a graph.

2.3 Modeling Data Using RDF Graphs

The term "ontology" in the Semantic Web refers to modeling a system's knowledge. This is based on RDF, in which, the perception of the world is modeled as a graph. The OWL language, that builds on top of RDF, is covered next, in Sect. 2.4.2.

RDF (Resource Description Framework) (Schreiber and Raimond 2014) is a data model for the Web, a framework that allows expressing information about resources. The creation of RDF stems from the ambition of providing a common representation of web resources and is, in fact, the first attempt to this direction. The first publication of RDF took place in 1999. As such, RDF is considered the cornerstone of the Semantic Web. Among its most important features is that it can represent data from other data models, making it easy to integrate data from multiple heterogeneous sources.

The main idea behind RDF is based on modeling every resource with respect to its relations (properties) to other web resources. These relations form triples, in which the first term is typically referred to as *subject*, the second one as *property* and the third one as *object*. The result of this approach is the creation of RDF Statements that contain triples, in the following form:

(*resource, property, resource*), or (*subject, property, object*).

RDF models data as a *directed* and *labeled* graph. In an RDF graph, more than one edges between nodes are allowed, RDF nodes do not need to be connected to one another, and it is allowed to find circle paths in the graph.

An RDF graph can be illustrated graphically in the following way: For each triple (subject; property; object), add to the graph the nodes *subject* and *object* and the edge *property* as follows: $subject \overset{property}{\rightarrow} object$. In Fig. 2.1, a part of an RDF graph that contains a blank node is drawn.

Nodes in RDF can be resources (URIs or IRIs), literals or blank nodes. The difference between resources and literals is that literals are not subject to further

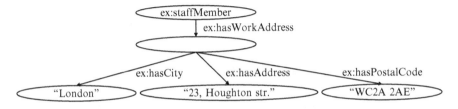

Fig. 2.1 RDF graph example

processing by RDF parsers. Contrarily, nodes that contain URI references (and blank nodes, otherwise known as bNodes) declare the presence of unique resources in the graph. However, blank nodes do not provide any extra information about these resources. Thus, blank nodes[2] are nodes in the graph that neither point somewhere nor can be referenced from outside the graph. In several RDF management implementations, local identifiers are assigned to blank nodes, these identifiers however are not persistent or portable. The blank node identifiers have a local scope and are merely artifacts of the RDF serialization.

Formally, an RDF graph is defined by the sets U (RDF URI references), B (blank nodes), and L (literals), with the sets U, B, and L being mutually disjoint among them. An RDF graph is a finite set of RDF triples. Formally, RDF triples are triples of the form

$$(v1; v2; v3) \in (U \cup B) \times U \times (U \cup B \cup L)$$

RDF serves as a foundation for a family of standards, including:

- RDF Schema: Documenting the meaning of RDF data (Sect. 2.3.3)
- OWL: Formalizing the meaning of RDF data (Sect. 2.4.2)
- SPARQL: Querying RDF data (Sect. 2.5)
- R2RML: Mapping relational data to RDF (Sect. 2.6)
- GRDDL: Mapping XML data to RDF (Sect. 2.7)

The latest version of RDF, RDF 1.1 is the first major update since the framework's standardization in 2004. Version 1.1 brings several changes to the standard, such as the introduction of *Named Graphs*, a concept adopted from SPARQL, and the introduction of the JSON-LD serialization syntax. Overall, RDF 1.1 was chartered to do some maintenance and fix some well-known issues.[3]

2.3.1 Using Namespaces

Before continuing with a more in-depth analysis of RDF graphs, we need to make special reference to namespaces, an idea widely used in the XML world. Namespaces were initially design to prevent confusion in XML names that originate from different sources. For instance, if an XML file contains an *example* element which needs to be differentiated from an *example* element belonging to another document—which happens often enough in XML files—each XML file must have its own namespace. This allows each *example* element to be uniquely identified. Therefore, if the two elements belong to different namespaces, namely *a* and *b*, respectively, then the elements can be referred to as *a:example* and *b:example*, thus making a

---

[2] Blank Nodes: www.w3.org/TR/2014/REC-rdf11-concepts-20140225/#section-blank-nodes.

[3] The full list of changes, along with all the technical details is described in www.w3.org/TR/rdf11-new/.

Table 2.1 Common namespaces in the Semantic Web

Prefix	Namespace URI (location)	Description
rdf	http://www.w3.org/1999/02/22-rdf-syntax-ns#	The built-in RDF vocabulary
rdfs	http://www.w3.org/2000/01/rdf-schema	RDFS puts an order to RDF (Sect. 2.3.3)
owl	http://www.w3.org/2002/07/owl	OWL terms (Sect. 2.4.2)
xsd	http://www.w3.org/2001/XMLSchema#	The RDF-compatible XML Schema datatypes
dc	http://purl.org/dc/elements/1.1/	The Dublin Core standard for digital object description (Sect. 2.8)
foaf	http://xmlns.com/foaf/0.1	The FOAF network (Sect. 2.8)
skos	http://www.w3.org/2004/02/skos/core#	Simple Knowledge Organization System (Sect. 2.8)
void	http://rdfs.org/ns/void#	Vocabulary of Interlinked Datasets (Sect. 2.8)
sioc	http://rdfs.org/sioc/ns#	Semantically-Interlinked Online Communities (Sect. 2.8)
cc	http://creativecommons.org/ns	Creative commons helps expressing licensing information
rdfa	http://www.w3.org/ns/rdfa	RDFa (Sect. 2.3.2)

clear distinction between the two. Elements in XML files can belong to any number of namespaces.

Namespaces are declared in the beginning of XML documents as follows:

```
xmlns:prefix="location".
```

Several namespaces are common in the Semantic Web, some of which are mentioned in Table 2.1 and presented in more detail in Sect. 2.8.

Finally, in this work, we use the *ex* namespace in order to form examples, for instance: *xmlns:ex="http://www.example.org/"*. Therefore, a resource in the *ex* namespace could be: *http://www.example.org#phd_student*, or *ex:phd_student*. Resources in the latter, shorter form are referred to as QNames.

2.3.2 RDF Serialization

So far, we have given a visual representation of an RDF graph, which makes it more understandable to the human reader. In practice though, RDF is mainly destined for machine consumption and for this reason, several ways to express RDF graphs in machine-readable formats have been proposed. The most prevalent among those formats, also called *serialization* formats, are:

- N-Triples (Carothers and Seaborne 2014a)
- Turtle (Prud'hommeaux and Carothers 2014)
- N-Quads (Carothers 2014)

- TriG (Carothers and Seaborne 2014b)
- RDF/XML (Gandon and Schreiber 2014)
- JSON-LD (Sporny et al. 2014)
- RDFa (Herman et al. 2013)

The first four languages are very similar belonging to the so-called Turtle family of RDF syntaxes. **N-Triples** is a simple serialization format that expresses each RDF triple in a single line ending with a period, as shown next:

```
1    <http://www.example.org/bradPitt>
     <http://www.example.org/isFatherOf>
     <http://www.example.org/maddoxJoliePitt>.
2    <http://www.example.org/bradPitt>
     <http://xmlns.com/foaf/0.1/name> "Brad Pitt".
3    <http://www.example.org/bradPitt>
     <http://xmlns.com/foaf/0.1/based_near> :_x.
4    :_x <http://www.w3.org/2003/01/geo/wgs84_pos#lat> "34.1000".
5    :_x <http://www.w3.org/2003/01/geo/wgs84_pos#long> "118.3333".
```

N-Triples is best suited for line-processing tools that consume RDF, since all the components of a triple can be found in one line. On the contrary, **Turtle** is an extension of N-Triples that reduces the verbosity of N-Triples by allowing namespace prefixes and removing multiple occurrence of the same resource in a set of triples sharing the same subject. For example, using the appropriate prefix mappings, the previous set of triples will be expressed in Turtle notation as follows:

```
1    PREFIX ex: <http://www.example.org/>.
2    PREFIX foaf: <http://xmlns.com/foaf/0.1/>.
3    PREFIX geo: <http://www.w3.org/2003/01/geo/wgs84_pos#>.
4
5    ex:bradPitt ex:isFatherOf ex:maddoxJoliePitt;
6                foaf:name "Brad Pitt";
7                foaf:based_near :_x.
8    :_x geo:lat "34.1000";
9        geo:long "118.3333".
```

The above set of triples is logically equivalent to the previous triples expressed in the N-Triples syntax. Lines 1–3 define the namespaces and the respective prefixes that abbreviate the URIs used. The triples shown in lines 5–7 all share the same subject which is only written down for the first triple. A semicolon at the end of a line specifies that the predicate-object pair that follows is connected to the most recently mentioned subject. The same rule applies to lines 8–9, which both have as subject a *blank node*. Blank nodes in Turtle are denoted following a syntax like :_x does.

N-Triples and Turtle do not allow for the specification of the named graph a triple may belong to. This deficiency is overcome by the N-Quads and TriG syntaxes, which are extensions of N-Triples and Turtle respectively and support the expression of so called *RDF datasets*. **N-Quads** simply enhances every RDF triple with a fourth element, which denotes the named graph that contains this triple. **TriG** follows the same rationale as Turtle, using the keyword GRAPH to specify named graphs, as in the following set of *quads*:

```
1    PREFIX ex: <http://www.example.org/>.
2    PREFIX foaf: <http://xmlns.com/foaf/0.1/>.
3    PREFIX geo: <http://www.w3.org/2003/01/geo/wgs84_pos#>.
4
5    GRAPH <http://www.example.org/graphs/brad> {
6    ex:bradPitt ex:isFatherOf ex:maddoxJoliePitt;
7                foaf:name "Brad Pitt";
8                foaf:based_near :_x.
9    :_x geo:lat "34.1000";
10       geo:long "118.3333".
11   }
```

These four text-based RDF syntaxes bear a lot of resemblance with each other and are the most prevalent ones, since they are more readable to the human eye. Interestingly though, the first proposed RDF syntax was **RDF/XML**, which is XML-based and more verbose than the four languages of the Turtle family. An RDF graph expressed in RDF/XML must be a valid XML file with all RDF triples enclosed in an <rdf:RDF> element. The five triples used in the previous examples would be expressed in RDF/XML as follows:

```
1    <?xml version="1.0" encoding="utf-8"?>
2    <rdf:RDF xmlns:ex="http://www.example.org/"
3    xmlns:foaf="http://xmlns.com/foaf/0.1/"
4    xmlns:geo="http://www.w3.org/2003/01/geo/wgs84_pos#">
5        <rdf:Description rdf:about="
     http://www.example.org/bradPitt">
6            <ex:isFatherOf rdf:resource="
     http://www.example.org/maddoxJoliePitt"/>
7            <foaf:name>Brad Pitt</foaf:name>
8            <foaf:based_near rdf:nodeID="A0"/>
9        </rdf:Description>
10       <rdf:Description rdf:nodeID="A0">
11           <geo:lat>34.1000</geo:lat>
12           <geo:long>118.3333</geo:long>
13       </rdf:Description>
14   </rdf:RDF>
```

The outer <rdf:RDF> element (lines 2–4) contains the namespace prefixes that will be used throughout the RDF graph. The subject of a triple is specified by the <rdf:Description> element and its URI is given as value of the rdf:about attribute. The predicate-object pairs that correspond to a given subject are represented as nested elements inside the <rdf:Description> element and are named after the predicate. This is the case for elements <ex:isFatherOf>, <foaf:name> and <foaf:based_near> (lines 6–8), each corresponding to a predicate. If the object of the triple is a resource, it is specified as the value of the rdf:resource attribute (line 6), while if it is a literal, it is represented as the content of the respective element (line 7). Blank nodes are specified via the rdf:nodeID attribute (line 8) and they can be used in other triples (lines 10–13), just like named nodes.

Another RDF syntax that is based on a popular web technology is **JSON-LD**, which was devised in order to make RDF even more accessible to web developers and processable by tools and software that can use JSON. An RDF graph expressed in JSON-LD is a valid JSON document, in which keys can be identified by a URI and the types of values can be explicitly specified. The example RDF graph used before is equivalent to the following JSON-LD document:

```
1    {
2      "@context":{
3        "foaf": "http://xmlns.com/foaf/0.1/",
4        "child": {
5          "@id": "http://www.example.org/isFatherOf",
6          "@type": "@id"
7        },
8        "name": "foaf:name",
9        "location": "foaf:based_near",
10       "geo": "http://www.w3.org/2003/01/geo/wgs84_pos#",
11       "lat": "geo:lat",
12       "long": "geo:long"
13     },
14     "@id": "http://www.example.org/bradPitt",
15     "child": "http://www.example.org/maddoxJoliePitt",
16     "name": "Brad Pitt",
17     "location": {
18       "lat": "34.1000",
19       "long": "118.3333"
20     }
21   }
```

The @id key (line 14) denotes the resource that is described by the JSON document and constitutes the subject of all RDF triples implied by it. Lines 15–20 contain key-value pairs for the resource ex:bradPitt and more specifically, his child, his name and the location of his residence. However, these key-value pairs do not contain the entire information needed for the construction of an RDF graph.

This information is provided by the @context element (lines 2–13), which specifies the needed namespace prefixes (for example, lines 3 and 10) as well as the mappings among keys and property URIs (for example, lines 4–7, 8, 9, 11 and 12). Furthermore, in case the object of a property should be considered a resource URI instead of a plain literal, this can be defined via the @type key, as shown in lines 4–7. These line essentially specify that the value of the child key (i.e. ex:maddoxJoliePitt in our case) will be a resource.

Finally, an RDF syntax that is quite different from the others, since it does not only represent an RDF graph but also combines it with HTML content is **RDFa**. RDFa introduces a number of HTML attributes that can be used to *annotate semantically* the content of an XHTML or HTML5 document. For example, let us consider the following XHTML code:

```
1    <html>
2    <head>
3       …
4    </head>
5    <body vocab="http://xmlns.com/foaf/0.1/">
6       <div resource="http://www.example.org/bradPitt">
7          <p>Famous American actor <span property="name">Brad
         Pitt</span> eldest son is
8        <a property="http://www.example.org/isFatherOf"
         href="http://www.example.org/maddoxJoliePitt">Maddox Jolie-
         Pitt</a>.
9          </p>
10       </div>
11    </body>
12    </html>
```

The vocab attribute (line 5) provides a URI that will be used as a base for all relative URI references that are mentioned within the body element. The resource attribute of the div element (line 6) specifies the URI which will be the subject of the triples that will be generated by the RDFa annotations that exist within this element. The property attribute specifies the absolute or relative URI reference of an RDF property. The value of that property will be either the content of that element (line 7) or the value of a resource or href attribute in the same HTML element (line 8). In the first case, the triple

```
ex:bradPitt foaf:name "Brad Pitt".
```

will be generated by an RDFa processor, while in the second case, the triple

```
ex:bradPitt ex:isFatherOf ex:maddoxJoliePitt.
```

is produced.

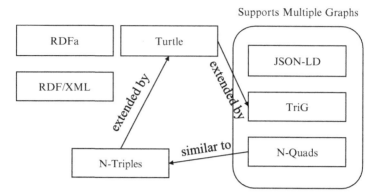

Fig. 2.2 Serialization formats in RDF 1.1 (*source*: www.w3.org/TR/rdf11-new/)

Figure 2.2 illustrates the relations among the various serialization formats defined in RDF 1.1.

This concludes a brief description of the various RDF syntaxes, which hopefully showcases the main differences among them. For a complete and normative reference, the interested reader can follow the official W3C specifications, which are referred to in the beginning of this section.

2.3.3 The RDF Schema

As stated previously in Sect. 2.3.2, RDF is simply a graph model that provides the basic constructs for defining a graph structure. An RDF graph can describe virtually anything, from real world entities to abstract concepts, as long as they can be modeled in a graph structure. The semantics of the RDF model do not specify what the nodes and edges of an RDF graph may represent. This is the purpose of so-called knowledge representation languages, which provide the necessary building blocks for the definition of domain vocabularies, i.e. sets of concept and relationships that pertain to a given domain. Perhaps the simplest such knowledge representation language is RDF Schema or RDFS for short (Brickley and Guha 2014).

RDFS is also said to be a semantic extension of RDF in the sense that it provides mechanisms for assigning meaning to the nodes and edges (i.e. resources and their properties) of an RDF graph. One common misconception that is often found among RDF newcomers states that RDF Schema and RDF have a similar relationship as XML Schema and XML do. Actually, this claim is not valid, since XML Schema imposes a specific structure that an XML document must follow in order to be valid, while RDF Schema essentially complements an RDF graph with concrete meaning.

The main features that RDFS introduces for enhancing an RDF graph with additional meaning are:

- Definition of groups of resources (called *classes*)
- Definition of hierarchies of classes
- Definition of hierarchies of properties
- Specification of application rules (i.e. *domain* and *range*) for properties

Examining those features one by one, we begin with the ability to group individual entities of the same type in *classes*. The members of a class are called *instances* of this class. This is one of the most fundamental functionalities for an ontology language, allowing a group of individual entities to share the same set of attributes and restrictions, in much the same way as classes in object-oriented programming languages do. In RDFS, a class is defined itself as an instance of the general rdfs:Class meta-class, while the membership of an instance in a class is stated via the rdf:type property. RDFS uses the namespace http://www.w3.org/2000/01/rdf-schema#, which is usually referred to as the abbreviated prefix rdfs, while the RDFS specification also refers to the RDF namespace http://www.w3.org/1999/02/22-rdf-syntax-ns#, often abbreviated as rdf, as shown in the previous notation. For example, the following RDF triple defines a new RDFS class ex:Actor, where the ex prefix refers to the example namespace http://www.example.org/:

```
ex:Actor rdf:type rdfs:Class.
```

Membership of the ex:bradPitt individual in the ex:Actor class is asserted via the following triple:

```
ex:bradPitt rdf:type ex:Actor.
```

If we recall the discussion on the formal definition of ontologies and knowledge bases (Sect. 1.2.7), we could say that the first triple represents intensional knowledge about the movie industry domain, while the second triple provides extensional knowledge, given the fact that it refers to a specific individual entity.

After the definition of classes, an ontology modeler usually defines relationships among those classes, in order to better capture the semantics of a particular domain. Containment relationships or subclass/superclass relationships are perhaps the most frequently used type of relationship that can be established among classes and this is the reason why even the simplest ontology languages have constructs for defining class hierarchies. In RDFS, this feature is provided via the rdfs:subClassOf property, which leads to interesting inferences that capture the intuition of a human user that is familiar with the domain under description. For example, the following triple establishes class ex:Actor as a subclass of the ex:MovieStaff class:

```
ex:Actor rdfs:subClassOf ex:MovieStaff.
```

The above ontology axiom entails that every instance of the ex:Actor class is also a member of the ex:MovieStaff class. Therefore, an inference service could also generate the following triple, according to the previous explicitly asserted triples:

```
ex:bradPitt rdf:type ex:MovieStaff.
```

The last triple is an inferred triple, i.e. a triple that is extracted from a systematic process (reasoning) based on present ontology axioms and already asserted facts about the individuals of the knowledge base. On the practical side, a knowledge system engineer has the choice of either storing all possible inferred triples or generating them on request of a calling user or application. Every strategy has its own issues that should be dealt with; the choice of which one to follow should take into account the efficiency of each strategy as well as the storage and response time tradeoff. Nevertheless, the choice of implementation for a semantics-aware system is out of the scope of the RDFS and OWL (to be introduced next in Sect. 2.4.2) standards, which just specify the rules under which new triples can be inferred from existing ones.

RDFS does not impose restrictions on the form of the class hierarchy that can be defined in an ontology. In other words, an RDFS ontology does not need to have a strict tree-like hierarchy as regular taxonomies do, but more complex graph-like structures are allowed, where each class may have more than one superclass.

RDFS also allows hierarchies of properties to be defined, just like class hierarchies. Property hierarchies are introduced via the rdfs:subPropertyOf property and allow the inference of additional triples that link a pair of RDF resources via more than one property. For example, the following triples define the ex:participatesIn and ex:starsIn properties and specify that the latter is a subproperty of the former:

```
ex:participatesIn rdf:type rdf:Property.
```

```
ex:starsIn rdf:type rdf:Property.
```

```
ex:starsIn rdfs:subPropertyOf ex:participatesIn.
```

The latter axiom essentially denotes that when two class instances are related via the ex:starsIn property, they should also be related via the ex:participatesIn property. This means that the following asserted triple:

```
ex:bradPitt ex:starsIn ex:worldWarZ.
```

combined with the previous subproperty axiom gives rise to the next inferred triple:

```
ex:bradPitt ex:participatesIn ex:worldWarZ.
```

Another fundamental feature of RDFS is the ability to define the classes to which RDF properties can be applied. This is one core difference of the RDF model when compared to other models, where properties or attributes are defined in the context of specific entities. In RDF, properties are defined and considered as stand-alone elements that can exist regardless of any classes or instances. This is why RDFS includes mechanisms for defining the class that a property can be applied to and also the class that constitutes the target of that property. Those mechanisms are provided via the `rdfs:domain` and `rdfs:range` properties respectively. When an RDFS class C is denoted as the domain of a property P, this means that any resource that occurs in the subject position of a triple containing P in the predicate position is an instance of type C. More simply put, any resource that has property P is an instance of class C. For example, let the following domain axiom hold:

```
ex:starsIn rdfs:domain ex:Actor.
```

Then, the following assertion:

```
ex:georgeClooney ex:starsIn ex:idesOfMarch.
```

combined with the previous domain axiom, entails the following inferred triple:

```
ex:georgeClooney rdf:type ex:Actor.
```

The previous example essentially states that, since only instances of the `ex:Actor` class can have the `ex:starsIn` property, the instance `ex:georgeClooney` must be an instance of the former class. A range axiom, on the other hand, allows an inference engine to infer that the value of a specific property is a member of a given class. Similarly to the previous examples, if we consider the next range axiom:

```
ex:starsIn rdfs:range ex:Movie.
```

an inference engine that takes into account the triples that have been previously mentioned in this section will produce the following triples:

```
ex:worldWarZ rdf:type ex:Movie.
```

```
ex:idesOfMarch rdf:type ex:Movie.
```

The role that domain and range axioms play in an RDFS ontology may often be misleading, especially for those who are new to knowledge representation formalisms and the so-called open world assumption (see Sect. 1.2.7.1). Contrarily to what most people expect when first encountering domain and range axioms, those do not function as constraints that data need to follow, but as rules that lead to the production of new triples and knowledge. For example, consider the following triple:

```
ex:georgeClooney ex:starsIn ex:bradPitt.
```

which obviously does not reflect the intention of the knowledge engineer, since it goes against human intuition. While one would have expected that an RDFS inference engine would be able to raise an error or warning, it will in fact infer the additional triple:

```
ex:bradPitt rdf:type ex:Movie.
```

based on the previously introduced range axiom. The above triple does not affect the consistency of the overall knowledge base, since there is not any RDFS restriction that prohibits an individual entity to be a member of both `ex:Actor` and `ex:Movie` classes. Such restrictions are met in more expressive ontology languages, such as OWL.

Reification (or reified statements) is another interesting mechanism in RDF. It is the ability to treat an RDF statement as an RDF resource, and hence to make assertions about that statement. In other words apart from "pure" statements that apply for describing web resources, RDF can also be used to create statements about other RDF statements, known as high-order statements. For instance, using reification, the following statement (in Turtle syntax):

```
<http://www.example.org/person/1> foaf:name "Brad Pitt ".
```

becomes:

```
<http://www.example.org/statement/5> a rdf:Statement;

rdf:subject <http://www.example.org/person/1>;

rdf:predicate foaf:name;

rdf:object "Brad Pitt".
```

The reification mechanism can be useful when referring to an RDF triple in order to describe properties that apply to it as a whole, e.g. to denote its provenance or assign a trust level to it. For example, the following triple assigns a trust level of 0.8 to the previous reified statement:

```
<http://www.example.org/statement/5> ex:hasTrust "0.8"^^xsd:float.
```

The above set of mechanisms, together with other constructs such as containers, makes RDFS a knowledge representation language of minimal expressiveness that allows for the definition of simple terminological knowledge that usually suffices for simple application scenarios. Table 2.2 gathers all the RDFS vocabulary. For representation of more complex domains though, in cases when RDFS expressiveness does not suffice, one has to resort to more expressive languages, such as OWL.

2.4   Ontologies Based on Description Logics

In the previous section we described how data can be modeled as a graph and now we will explain how knowledge can be modeled using an ontology language whose roots are in Description Logics (DL).

2.4.1   Description Logics

The effort in providing a theoretical ground for logic led to the creation of the OWL language. The first attempts of Semantic Web languages include DAML, OIL, and DAML+OIL. The latest results in this direction are OWL and OWL 2, which attempt to model knowledge for the Semantic Web. These languages are characterized by formal semantics, while syntactically being compatible to the RDF serializations. Therefore, ontologies in OWL can be queried in the same approach as in RDF graphs (for querying, see Sect. 2.5).

Table 2.2 The RDFS vocabulary

Declaration	Explanation
Classes	
`rdfs:Resource`	All things described by RDF are instances of the class `rdfs:Resource`. All other classes are subclasses of this class, which is an instance of `rdfs:Class`
`rdfs:Literal`	The class of all literals. It is an instance of `rdfs:Class`. Literals are represented as strings but they can be of any XSD datatype
`rdfs:langString`	The class of language-tagged string values. It is a subclass of `rdfs:Literal` and an instance of `rdfs:Datatype`. Example: "foo"@en
`rdfs:Class`	The class of all classes, i.e. the class of all resources that are RDF classes
`rdfs:Datatype`	The class of all the data types. The `rdfs:Datatype` is in the same time an instance but also a subclass of `rdfs:Class`. Every instance of `rdfs:Datatype` is also a subclass of `rdfs:Literal`
`rdf:HTML`	The class of HTML literal values. It is an instance of `rdfs:Datatype` and a subclass of `rdfs:Literal`
`rdf:XMLLiteral`	The class of XML literal values. It is an instance of `rdfs:Datatype` and a subclass of `rdfs:Literal`
`rdf:Property`	The class of all RDF properties
Properties	
`rdfs:domain`	Declares the domain of a property P, that is the class of all the resources that can appear as S in a triple (S, P, O)
`rdfs:range`	Declares the range of a property P, that is the class of all the resources that can appear as O in a triple (S, P, O)
`rdf:type`	A property that is used to state that a resource is an instance of a class
`rdfs:label`	A property that provides a human-readable version of a resource's name
`rdfs:comment`	A property that provides a human-readable description of a resource
`rdfs:subClassOf`	Corresponds a class to one of its superclasses. Note that a class can have more than one superclasses
`rdfs:subPropertyOf`	Corresponds a property to one of its superproperties. A property can have more than one superproperties
Container classes and properties	
`rdfs:Container`	A superclass of all classes that can contain instances such as `rdf:Bag`, `rdf:Seq` and `rdf:Alt`
`rdfs:member`	It is a superproperty to all the properties that declare that a resource belongs to a class that can contain instances (container)
`rdfs:ContainerMembership Property`	A property that is used in declaring that a resource is a member of a container. Every instance is a subproperty of `rdfs:member`

(continued)

Table 2.2 (continued)

Declaration	Explanation
rdf:Bag	The class of unordered containers. It is a subclass of rdfs:Container
rdf:Seq	The class of ordered containers. It is a subclass of rdfs:Container
rdf:Alt	The class of containers of alternatives. It is a subclass of rdfs:Container
Collections	
rdf:List	The class of RDF Lists. An instance of rdfs:Class that can be used to build descriptions of lists and other list-like structures
rdf:nil	An instance of rdf:List that is an empty rdf:List
rdf:first	The first item in the subject RDF list. An instance of rdf:Property
rdf:rest	The rest of the subject RDF list after the first item. An instance of rdf:Property
rdf:_1, rdf:_2, rdf:_3, etc.	All are both a sub-property of rdfs:member and an instance of the class rdfs:ContainerMembership Property
Reification vocabulary	
rdf:Statement	The class of RDF statements. An instance of rdfs:Class, intended to represent the class of RDF statements
rdf:subject	The subject of the subject RDF statement. It is an instance of rdf:Property, used to state the subject of a statement
rdf:predicate	The predicate of the subject RDF statement. An instance of rdf:Property, used to state the predicate of a statement
rdf:object	The object of the subject RDF statement. An instance of rdf:Property, used to state the object of a statement
Utility properties	
rdf:value	Idiomatic property used for describing structured values. It is an instance of rdf:Property
rdfs:seeAlso	A property that indicates a resource that might provide additional information about the subject resource
rdfs:isDefinedBy	A property that indicates a resource defining the subject resource. For instance, the defining resource may be an RDF vocabulary in which the subject resource is described

(continued)

DL offers the language for the description and manipulation of independent individuals, roles, and concepts (Brachman and Schmolze 1985). There can be many DL languages, the difference from one language to another being the language semantics. These languages are used in order to describe the world using formulas. In general, the formulas are constructed using sets of concepts, roles, individuals, and constructors (such as intersection $\wedge$, union $\vee$, exists $\exists$, and for each

∀). Variables in DL can represent arbitrary world objects. For instance, a Description Logic formula can declare that a number x, which is greater than zero exists: $\exists x : greaterThan(x; 0)$. By adding more constructors to the basic DL language it is possible to increase its expressiveness in order to describe more complex concepts. Therefore, the expressiveness used in describing the world varies according to the subset of the language that is used.

For instance, adding the concept conjunction capability, thus allowing declarations such as $(C \cup D)$ that enable description of concepts such as $Parent = Father \cup Mother$, we obtain a more descriptive DL language. From a semantics perspective, these languages are not different from each another. The differentiation in the occurring languages affects not only the world description capabilities but also the behavior and performance of the algorithms that process them. The OWL language is directly related to DL, since the different "flavors" correspond to different DL language subsets, as we analyze in more detail in the next section.

2.4.2 The Web Ontology Language (OWL)

OWL stands for Web Ontology Language. It is the successor of DAML + OIL and it is the current recommendation by W3C, currently in version 2. The creation of the OWL language (McGuinness and van Harmelen 2004) officially begins with the initiation of the DAML project. DAML, combined to OIL led to the creation of DAML + OIL which was an extension of RDFS. OWL is the successor of DAML + OIL. The language's roots are in DL (Horrocks et al. 2003), based on standard predicate calculus. As such, the world in OWL is viewed as sets of classes, properties and individuals.

The OWL language was designed in order to allow applications to process the information content itself instead of simply presenting the information. According to W3C, the goal is to provide a schema that will be compatible both to the World Wide Web architecture and the Semantic Web. Therefore, encoding ontologies in the OWL language makes information more processable both by machines and humans.

The first OWL version comprised three flavors: Lite, DL and Full. OWL Lite was designed keeping in mind that it had to resemble RDFS. OWL DL was a more interesting subset of Description Logics: OWL DL guaranteed that all reasoning procedures are finite and return a result. OWL Full held the whole wealth and expressiveness of the language. This means that more expressive sentences can be expressed in OWL Full. However, even for small declaration sets in OWL Full, reasoning is not guaranteed that it will be finite.

An OWL ontology may also comprise declarations from RDF and RDFS, such as `rdfs:subClassOf`, `rdfs:range` and `rdf:resource`. OWL uses them and relies on them in order to model its concepts. Figure 2.3 illustrates how OWL is based on RDF and RDFS for the definition of its vocabulary.[4]

---

[4]A full reference to the OWL 2 language can be found at www.w3.org/TR/owl-quick-reference/.

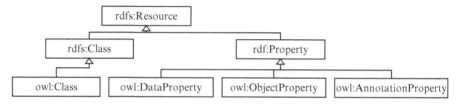

Fig. 2.3 OWL relies on RDF(S) for the definition of its concepts

The OWL 2 language is the latest version of the language, as evolved from OWL. OWL 2 has a very similar overall structure to OWL 1 and it is backwards compatible in the sense that each OWL 1 ontology is also an OWL 2 ontology.

OWL 2, while remaining compatible with OWL, presents a set of additional features, introducing new functionality with respect to OWL 1. These features have been requested by users, and are features for which effective reasoning algorithms are available, and that OWL tool developers are willing to support. These new features include:

- Additional property and qualified cardinality constructors, in order to allow for the assertion of minimum, maximum or exact qualified cardinality restrictions, object and data properties (e.g. `ObjectMinCardinality`, `DataMaxCardinality`).
- Property chains. In OWL, the means were not provided to define properties as a composition of other properties; hence, it was not possible to propagate a property (e.g. `locatedIn`) along another property (e.g. `partOf`), something allowed in OWL 2 with the usage of constructs `ObjectPropertyChain` in a `SubObjectPropertyOf` axiom.
- Extended datatype support. While the first version of OWL provided support for only integers and strings as datatypes, it did not support the definition of subsets of these datatypes. For example, one could state that every person has an age, which is of type integer, but could not restrict the range of that datatype value.
- Simple metamodeling. OWL DL imposed a strict separation between the names of, e.g. classes and individuals. In OWL 2, this separation is somewhat relaxed in order to allow different uses of the same term.
- Extended annotations. OWL 1 allowed annotations, such as a label or a comment, to be given for each ontology entity. However, annotations of axioms were not allowed, e.g. who asserted an axiom or when, something that is present in OWL 2.
- Extra syntactic sugar: In order to make common patterns easier to write, without changing the expressiveness, semantics, or complexity of the language.

OWL 2 also defines three new flavors (Profiles), based on completely new principles, and a new syntax, the OWL 2 Manchester Syntax. The OWL 2 Profiles are sub-languages (syntactic subsets, in the sense that was described in Sect. 2.4.1) of OWL 2 that demonstrate better abilities under particular application scenarios (Hitzler et al. 2012):

- **OWL 2 EL** enables polynomial time algorithms for all the standard reasoning tasks; it is particularly suitable for applications where very large ontologies are needed, and where expressive power can be traded for performance guarantees.

```
Class: VegetarianPizza
EquivalentTo:
Pizza and
not (hasTopping some FishTopping) and
not (hasTopping some MeatTopping)

DisjointWith:
NonVegetarianPizza
```

Listing 2.1 Manchester syntax example

- **OWL 2 QL** enables conjunctive queries to be answered in LOGSPACE (more precisely, AC^0) using standard relational database technology; it is particularly suitable for applications where relatively lightweight ontologies are used to organize large numbers of individuals and where it is useful or necessary to access the data directly via relational queries (e.g. SQL).
- **OWL 2 RL** enables the implementation of polynomial time reasoning algorithms using rule-extended database technologies operating directly on RDF triples; it is particularly suitable for applications where relatively lightweight ontologies are used to organize large numbers of individuals and where it is useful or necessary to operate directly on data in the form of RDF triples.

The Manchester OWL syntax is designed in order to offer a user-friendly syntax for OWL 2 descriptions, but the syntax can also be used in order to write entire OWL 2 ontologies. The following example represents the full class description of Vegetarian Pizzas (Horridge et al. 2006) (Listing 2.1).

2.5   Querying the Semantic Web with SPARQL

Representation of knowledge in feature-rich machine readable formats is crucial in the Semantic Web, as is the ability to query and search effectively this knowledge. The latter is the role of SPARQL (SPARQL Protocol and RDF Query Language) (Harris and Seaborne 2013), a W3C recommendation that is essentially for RDF what SQL is for relational databases. The main component of the SPARQL protocol is the SPARQL query language, used for retrieving and restructuring information from RDF graphs. The SPARQL query language will be the main subject of this section. Other prominent members of the SPARQL family of recommendations include SPARQL Update, that is a language for updating and creating new RDF graphs, the SPARQL 1.1 Protocol, Graph Store HTTP Protocol for the transfer of queries to appropriate query endpoints, SPARQL 1.1 entailment regimes, that is a specification of the semantics of SPARQL queries under various entailment schemes, and specifications about the serialization of query results in JSON, CSV and TSV, and XML formats.

SPARQL has reached a relatively mature state, incorporating a number of advanced features that most query languages should have, such as aggregates, nested subqueries, ordering and other solution modifiers. Furthermore, SPARQL supports different query forms, which allow for the construction of new RDF graphs (COSTRUCT queries) or can return a boolean value that denotes whether a given query has a solution or not (ASK queries).

The most popular type of SPARQL query, though, is the SELECT query, which returns a set of *bindings* of variables to RDF terms. A binding is simply a mapping from a variable to a URI or an RDF literal. A concept that is central to a SPARQL query is the *triple pattern*, which resembles an RDF triple with the only difference that it may also contains variables in one or more positions. An RDF triple *matches* a triple pattern if there is an appropriate substitution of variables with RDF terms that makes the triple pattern and the RDF triple equivalent. For example, the triple

```
ex:bradPitt rdf:type ex:Actor.
```

matches the triple pattern

```
?x rdf:type ex:Actor
```

under the substitution $\theta = \{?x \mid ex : bradPitt\}$. A set of triple patterns build a *basic graph pattern*, which is the heart of a SPARQL SELECT query. For example, the following simple SPARQL query contains in its WHERE clause a basic graph pattern of two triple patterns:

```
SELECT ?x ?date
WHERE{
    ?x rdf:type ex:Actor.
    ?x ex:bornIn ?date
}
```

In order for a basic graph pattern to be matched by a set of RDF triples, there must exist a substitution of variables, such that *all* triple patterns are matched by an RDF triple. In the example query above, resources are sought that are of type ex:Actor *and* have a value for the ex:bornIn property. The SELECT keyword specifies the variables that will be returned. Therefore, the above query retrieves all actors that have a specified birth date and that date from an input RDF graph. Let us consider, for example, the following set of RDF triples:

```
ex:bradPitt rdf:type ex:Actor;
    ex:bornIn "1963"^^xsd:gYear.
ex:angelinaJolie rdf:type ex:Actor;
    ex:bornIn "1975"^^xsd:gYear.
ex:jenniferAniston rdf:type ex:Actor.
```

then the previous SPARQL query would have the following *solution*:

?x	?date
ex:bradPitt	"1963"^^xsd:gYear
ex:angelina Jolie	"1975"^^xsd:gYear

The resource `ex:jenniferAniston` is not included in the results, since there is not any RDF triple that matches the second triple pattern. In order to retrieve all actors and actresses and get their birthdate only if it is specified in the RDF graph, the `OPTIONAL` keyword would have to be used. The `OPTIONAL` keyword denotes optional basic graph patterns, as in the following example query:

```
SELECT ?x ?date

WHERE{

  ?x rdf:type ex:Actor.

  OPTIONAL{?x ex:bornIn ?date}

}
```

The above query would generate the following solution:

?x	?date
ex:bradPitt	"1963"^^xsd:gYear
ex:angelinaJolie	"1975"^^xsd:gYear
ex:jenniferAniston	

where the resource `ex:jenniferAniston` is included in the results, but variable `?date` is left unbound for that specific binding. In order to specify alternative graph patterns, the `UNION` keyword is used:

```
SELECT ?x

WHERE{

  {?x rdf:type ex:Actor}

  UNION

  {?x rdf:type ex:Director}

}
```

The above query searches for resources that are of type `ex:Actor` or `ex:Director`, including cases where a resource may belong to both classes. In case one wants to set a condition that limits the result of a SPARQL query, she can use the `FILTER` keyword:

```
SELECT ?x ?date

WHERE{

  ?x rdf:type ex:Actor.

  ?x ex:bornIn ?date.

  FILTER(?date < "1975"^^xsd:gYear)

}
```

The previous query returns all actors that were known to be born before 1975 and their corresponding birth date. Furthermore, SPARQL includes solution modifiers that order or limit the results of a SPARQL query with the use of the ORDER BY and `LIMIT` keywords. For example, the following query returns actors ordered by their date of birth, with the younger ones mentioned first:

```
SELECT ?x ?date
WHERE{
  ?x rdf:type ex:Actor.
  ?x ex:bornIn ?date.
 }
ORDER BY DESC(?date)
```

With the use of the LIMIT keyword, one can get the ten youngest actors in an RDF graph:

```
SELECT ?x ?date
WHERE{
  ?x rdf:type ex:Actor.
  ?x ex:bornIn ?date.
 }
ORDER BY DESC(?date)
LIMIT 10
```

We end this section by briefly presenting the rest of SPARQL query forms. A CONSTRUCT query can generate an RDF graph based on a graph template using the solutions of a SELECT query. For example, the following CONSTRUCT query generates a new RDF triple for every actor that has a salary larger than 1 million euros, assigning him to the ex:HighlyPaidActor class.

```
CONSTRUCT {?x rdf:type ex:HighlyPaidActor}
WHERE{
  ?x rdf:type ex:Actor.
  ?x ex:hasSalary ?salary.
  FILTER(?salary > 1000000)
 }
```

An ASK query responds with a boolean value that specifies whether a graph pattern is satisfied by a subgraph of the considered RDF graph. For example, the following ASK query returns true if the RDF graph contains an actor named "Cate Blanchett":

```
ASK {
  ?x rdf:type ex:Actor.
  ?x foaf:name "Cate Blanchett".
 }
```

Finally, a DESCRIBE query returns a set of RDF statements that contain data about a given resource. The exact structure of the returned RDF graph depends on the implementation choices of the specific SPARQL engine.

```
DESCRIBE <http://www.example.org/bradPitt>.
```

For a complete presentation of the various SPARQL features, see (Harris and Seaborne 2013).

2.6 Mapping Relational Data to RDF

It has been argued that one of the reasons for the slow uptake of the Semantic Web vision is the lack of sufficient volume of RDF data, which will act as the fuel for the development of innovative and smart tools and applications (Hendler 2008). Since a large part of available data, existing not only on the Web but mostly outside of it, is stored in relational databases, the research community has been coming up with methods and tools for the translation of relational data to RDF. As we will see in Sect. 4.2, a fair number of tools are available for this task, each one with its own set of features and, unfortunately, its own mapping language. The need for free reuse of relational-to-RDF mappings (otherwise known as RDB-to-RDF mappings) among the wide array of offered tools led to the creation of R2RML (Das et al. 2012) by W3C, as a standard formalism for expressing such mappings.

R2RML stands for "RDB to RDF Mapping Language" and provides a vocabulary that allows for the definition of RDF views over a given relational schema. Since R2RML is database vendor-agnostic, a mapping over a specific relational schema can be reused among various database management systems, provided that it does not contain any proprietary SQL features. A mapping in R2RML is itself represented as an RDF graph following the Turtle syntax, called *R2RML mapping graph*.

In a nutshell, an R2RML mapping associates relational views with functions that generate RDF terms. These relational views are called *logical tables* and can be either relations or views defined in the given relational schema or custom views over that schema. The RDF generating functions are called *term maps* and are distin-guished according to the position of the RDF term in the generated triple, i.e. to *subject, predicate,* and *object maps*. The necessary term maps are appropriately grouped to form a *triples map*, a function that maps relational data to a set of RDF triples. An R2RML mapping can contain one or more triples maps. Furthermore, R2RML offers the ability to organize generated RDF triples into named graphs by defining respective *graph maps*.

Let us use a concrete example in order to better convey the essence of the R2RML standard. Suppose we want to export a part of the simple relational instance of Fig. 2.4 as an RDF graph. This relational instance contains data from the movie domain, correlating actors and directors with films they have participated in and, for the sake of simplicity, we assume that every film has at most one director. We also define an R2RML mapping, as shown in Listing 2.2.

The R2RML mapping graph of Listing 2.2 is the simplest possible, containing a single triples map that specifies the generation of a set of RDF triples for every row of the "FILM" relation. Every triples map must have exactly one logical table (line 6 in Listing 2.1), one subject map (lines 7–10) and one or more predicate-object maps (lines 11–19). A logical table essentially represents an SQL result set, each row of which gives rise to an RDF triple (or quad in the general case, when named graphs are used). A subject map is simply a term map that produces the subject of an RDF triple. Term maps can be categorized to *constant-valued, column-valued* and *template-valued* ones, depending on the rule that needs to be followed for the

<table>
<tr><td colspan="4" align="center">FILM</td></tr>
</table>

ID	Title	Year	Director
1	The Hunger Games	2012	1

DIRECTOR

ID	Name	Birth Year
1	Gary Ross	1956

ACTOR

ID	Name	Birth Year	Birth Location
1	Jennifer Lawrence	1990	Louisville,KY
2	Josh Hutcherson	1992	Union, KY

FILM2ACTOR

FilmID	ActorID
1	1
1	2

Fig. 2.4 Example relational instance

```
1    @prefix rr: <http://www.w3.org/ns/r2rml#>.
2    @prefix ex: <http://www.example.org/>.
3    @prefix dc: <http://purl.org/dc/terms/>.
4
5    <#TriplesMap1>
6        rr:logicalTable [ rr:tableName "FILM" ];
7        rr:subjectMap [
8            rr:template
     "http://data.example.org/film/{ID}";
9            rr:class ex:Movie;
10        ];
11        rr:predicateObjectMap [
12            rr:predicate dc:title;
13            rr:objectMap [ rr:column "Title" ];
14        ];
15        rr:predicateObjectMap [
16            rr:predicate ex:releasedIn;
17            rr:objectMap [ rr:column "Year";
18                           rr:datatype xsd:gYear;];
19        ].
```

Listing 2.2 Example of an R2RML triples map

generation of the RDF term. The subject map shown in Listing 2.6 is a template-valued one, since it uses a common template for all rows of the logical table (the "FILM" relation in that case), where the difference lies in the value of the "ID" column for each row. It also specifies, via the `rr:class` property, that every generated resource will be an instance of the `ex:Movie` ontology class.

Predicate-object maps, in a triples map, are functions that generate pairs of predicates and objects for the subject generated by a subject map. Every predicate-object map must contain at least one predicate map and at least one object map, which generate the predicate and object respectively of an RDF triple. Both predicate-object

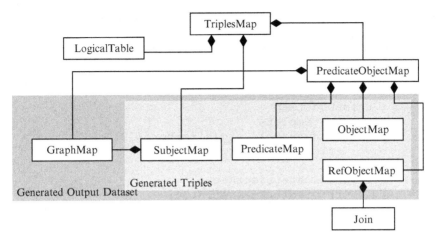

Fig. 2.5 Overview of R2RML constructs (*source*: www.w3.org/TR/r2rml/)

maps depicted in Listing 2.2 consist of a constant-valued predicate map, defined by the `rr:predicate` property (lines 12 and 16 respectively) and a column-valued object map indicated by the presence of the `rr:column` property (lines 13 and 17 respectively). Those maps essentially specify that the values of the "Title" column will be the values of the `dc:title` property and the values of the "Year" column will be the values of the `ex:releasedIn` property. The latter object map also specifies the datatype of the RDF literal that will be generated.

Therefore, the result of applying the triples map of Listing 2.2 to the relational instance shown in Fig. 2.5 will be the following set of triples in Turtle notation:

```
<http://data.example.org/film/1> a ex:Movie;

    dc:title "The Hunger Games";

    ex:releasedIn "2012"^^xsd:gYear.
```

It is important to note that R2RML allows the creation of RDF graphs that reuse terms from external ontologies, such as Dublin Core, which contains the `dc:title` property used in the previous triples map.

R2RML also supports the exploitation of foreign keys to link entities described in separate tables. For example, let us consider the following triples map that operates on the "DIRECTOR" relation and also add another predicate-object map in `TriplesMap1` introduced in Listing 2.2.

In a way similar to that of `TriplesMap1`, the `TriplesMap2` triples map of Listing 2.3 specifies the generation of three RDF triples per row of the "DIRECTOR" relation. For the case of the relational instance of Fig. 2.5, the following triples will be generated:

```
<http://data.example.org/director/1> a ex:Director;

    foaf:name "Gary Ross";

    ex:bornIn "1956"^^xsd:gYear.
```

```
1   @prefix rr: <http://www.w3.org/ns/r2rml#>.
2   @prefix ex: <http://www.example.org/>.
3   @prefix foaf: <http://xmlns.com/foaf/0.1/>.
4
5   <#TriplesMap2>
6       rr:logicalTable [ rr:tableName "DIRECTOR" ];
7       rr:subjectMap [
8           rr:template
    "http://data.example.org/director/{ID}";
9           rr:class ex:Director;
10      ];
11      rr:predicateObjectMap [
12          rr:predicate foaf:name;
13          rr:objectMap [ rr:column "Name" ];
14      ];
15      rr:predicateObjectMap [
16          rr:predicate ex:bornIn;
17          rr:objectMap [ rr:column "BirthYear";
18                         rr:datatype xsd:gYear; ];
19      ].
20
21  <#TriplesMap1>
22      rr:predicateObjectMap[
23          rr:predicate ex:directedBy;
24          rr:objectMap[
25                      rr:parentTriplesMap
    <#TriplesMap2>;
26                      rr:joinCondition[
27                              rr:child "Director";
28                              rr:parent "ID";
29                              ];
30                  ];
31      ].
```

Listing 2.3 Example of an R2RML referencing object map

The additional predicate-object map defined for `TriplesMap1` contains a *referencing object map*, which is responsible for the generation of the object of a triple. Referencing object maps are a special case of object maps that essentially follow the generation rules of the subject map of another triples map, called parent triples map. Join conditions are also specified in order to select the appropriate row of the logical table of the parent triples map. Without going into much detail, we observe that the object map of lines 24–30 in Listing 2.7 generates an RDF resource, as specified by the subject map of `TriplesMap2` (lines 7–10) by exploiting the foreign key relationship of "Director" and "ID" columns of "FILM" and "DIRECTOR" relations respectively. It is important to note that it is not necessary for this foreign key relationship to be explicitly defined as a constraint in the relational schema, since

R2RML is flexible enough to allow the linkage of any column set among different relations. For the current predicate-object map example, the following triple will be generated:

```
<http://data.example.org/film/1>

  ex:directedBy

<http://data.example.org/director/1>.
```

R2RML is also very flexible with respect to the definition of the logical table of a triples map, allowing the definition of custom views as in the example of Listing 2.4.

The R2RML view defined in line 6 via the rr:sqlQuery property could have alternatively been defined as a SQL view in the underlying database system. However, as it is not always feasible or desired for the mapping author to define new views in the relational schema, R2RML introduces the feature of R2RML views. R2RML

```
1   @prefix rr: <http://www.w3.org/ns/r2rml#>.
2   @prefix ex: <http://www.example.org/>.
3   @prefix dc: <http://purl.org/dc/terms/>.
4
5   <#TriplesMap3>
6       rr:logicalTable [ rr:sqlQuery """
                              SELECT ACTOR.ID AS ActorId,
    ACTOR.Name AS ActorName, ACTOR.BirthYear AS
    ActorBirth, FILM.ID AS FilmId FROM ACTOR, FILM,
    FILM2ACTOR WHERE FILM.ID=FILM2ACTOR.FilmID AND
    ACTOR.ID=FILM2ACTOR.ActorID; """];
7       rr:subjectMap [
8           rr:template
    "http://data.example.org/actor/{ActorId}";
9           rr:class ex:Actor;
10      ];
11      rr:predicateObjectMap [
12          rr:predicate foaf:name;
13          rr:objectMap [ rr:column "ActorName" ];
14      ];
15      rr:predicateObjectMap [
16          rr:predicate ex:bornIn;
17          rr:objectMap [ rr:column "ActorBirth";
18                         rr:datatype xsd:gYear;];
19      ].
20      rr:predicateObjectMap [
21          rr:predicate ex:starsIn;
22          rr:objectMap [ rr:template
    "http://data.example.org/film/{ID}";
23                         ];
24      ].
```

Listing 2.4 Example of a triples map with a custom R2RML view views must follow a number of rules (e.g. the produced result set must not contain columns with the same name) and they are used just as the native logical tables encountered so far. Therefore, the triples map defined in Listing 2.4 generates the following set of triples:

```
<http://data.example.org/actor/1> a ex:Actor;
    foaf:name "Jennifer Lawrence";
    ex:bornIn "1990"^^xsd:gYear;
    ex:starsIn
<http://data.example.org/film/1>.
<http://data.example.org/actor/2> a ex:Actor;
    foaf:name "Josh Hutcherson";
    ex:bornIn "1992"^^xsd:gYear;
    ex:starsIn
<http://data.example.org/film/1>.
```

R2RML also includes several more advanced features for the organization of generated triples in named graphs, the definition of blank nodes or the specification of a generated literal's language, but a complete presentation goes beyond the scope of the current section. A graphical sketch of the main R2RML constructs redrawn from the official W3C recommendation document, is shown in Fig. 2.5.

W3C has also specified a default strategy for translating a relational instance to an RDF graph. This transformation, which defines a standard URI generation policy for all kinds of relations, is called Direct Mapping (Arenas et al. 2012), maps every relation to a new ontology class and every relation row to an instance of the respective class. The strategy specified by Direct Mapping is a trivial one, since the generated RDF graph essentially mirrors the relational structure, but it can be useful as a starting point for mapping authors who can then augment and customize it accordingly by using the rich features of R2RML.

2.7 Other Technologies

Besides the main building blocks of the Semantic Web, which were presented in the previous sections and deal with the issues of describing, modeling and querying data, there are a number of other technologies that fill in the gaps and complement the Semantic Web ecosystem. These technologies are mainly developed by W3C and deal with, among others, the issues of defining rules as an additional means to encode data and providing standard ways for communicating with a Linked Data dataset:

- **POWDER (Protocol for Web Description Resources)**: POWDER (Scheppe 2009) is an XML-based protocol that allows for the provision of information about RDF resources. A POWDER document, besides its own provenance information, i.e. information about its creator, when it was created and so on, contains a list of so-called description resources. A description resource contains a set of property-value pairs, which may be expressed in RDF, for a given resource or group of resources. This mechanism allows for quick annotation of large amounts of content with metadata and is considered ideal for scenarios involving personalized delivery of content, trust control and semantic annotation.

- **RIF (Rule Interchange Format)**: Rules are an important component of the Semantic Web layered architecture, allowing for the derivation of statements that follow from a set of premises. Unlike the rest of the components of the Semantic Web stack, rules present a great amount of diversity, making impossible the definition of a single uniform rule language suitable for all purposes. This is the reason why W3C decided to define RIF (Rule Interchange Format) (Kifer and Boley 2010) as an extensible framework for the representation of various *rule dialects*. Therefore, RIF is essentially a family of specifications, currently defining three rule dialects: RIF-BLD (Basic Logic Dialect), RIF-Core and RIF-PRD (Production Rule Dialect). The first two are logic-based dialects, while the third one is a dialect that contains rules with actions, mainly used in production or business rule systems. RIF also includes RIF-FLD (Framework for Logic Dialects), a framework that allows the definition of other rule dialects, and also specifies the interface of rules expressed in a logic-based RIF dialect with RDF and OWL.

- **SPIN (SPARQL Inferencing Notation)**: Despite RIF being the official W3C recommendation as a rule authoring language, its uptake is not as massive as it was expected, mainly due to its complexity and the lack of supporting tools. Another approach for defining rules in the Semantic Web community is SPIN (Knublauch et al. 2011), which is a SPARQL-based language for the definition of production rules. SPIN allows for the linkage of RDFS and OWL classes with constraint checks and inference rules that define their expected behavior, in a manner similar to behavior of classes in object-oriented languages. These constraints and rules are expressed in SPARQL, making SPIN more accessible to Semantic Web programmers, and are linked to classes via SPIN's modeling vocabulary. Nevertheless, the future plans of W3C with respect to SPIN and whether it will ever be a W3C recommendation remains unknown.

- **GRDDL (Gleaning Resource Descriptions from Dialects of Languages)**: GRDDL (Connolly 2007) (pronounced "griddle") is a W3C recommendation that provides mechanisms that specify how an RDF representation of an XML document may be constructed. GRDDL can be applied to XHTML and XML documents and, with the use of appropriate markup attributes, defines XSLT transformations that, when applied to the original document, generate a semantically equivalent RDF/XML document. Therefore, GRDDL is the standard way of converting an XML document to RDF, although its use may appear daunting to users that are not familiar with XSLT.

- **LDP (Linked Data Platform)**: Linked Data Platform (Speicher et al. 2014) is the latest W3C effort in specifying best practices and guidelines for organizing and accessing *Linked Data Platform (LDP) resources*—which can be either RDF or non-RDF—via HTTP RESTful services. The Linked Data Platform recommendation specifies the expected behavior of a Linked Data server when it receives an HTTP request, allowing for the creation, deletion and update of resources. The specification also defines the concept of Linked Data Platform

Container, which serves as a collection of, usually homogeneous, resources. The LDP specification is expected to establish a standard behavior for all Linked Data systems. A notable example of a framework that implements LDP is Apache Marmotta (see Sect. 3.4.2.2).

2.8 Ontologies and Datasets

Besides the machine-consumable serializations of technologies such as RDF and OWL, several attempts at assigning meaning to the resources and their relationships exist in the bibliography, leading at the crystallization of these domain-specific concepts and their relationships into several ontologies that serve the purpose of providing common, shared descriptions of several domains.

Apart from the syntactical groundwork that is laid by the standards, popular vocabularies should be used in order for the described information to be commonly, unambiguously interpreted and understood. These vocabularies can be reused by various data producers when describing data about a given subject, making such data semantically interoperable. One should note the slight difference between such vocabularies and datasets, which, simply put, are collections of facts involving specific individual entities and ontologies, which provide an abstract, high-level, terminological description of a domain, containing axioms that hold for every individual.

A great number of ontologies exist nowadays, covering the majority of subject domains: from geographical and business vocabularies to life sciences, literature and media. Several ontology search engines and specialized directories exist for the discovery of appropriate vocabularies for a specific application domain. Notable examples include Schemapedia,[5] Watson,[6] Swoogle[7] and the Linked Open Vocabularies (LOV) registry.[8] The latter constitutes the most accurate and comprehensive source of ontologies used in the Linked Data cloud. Vocabularies in LOV are described by appropriate metadata and classified to domain spaces, while LOV also enables full text search at vocabulary and element level, enabling a domain modeler to discover the most relevant vocabulary for her case.

Next, we will take a look at some of the most widespread ontologies that are used in several datasets and establish a set of core vocabularies for the unambiguous description of knowledge. We will also mention examples of large and trusted datasets that constitute central hubs in the Linked Data global graph. Among the most popular vocabularies are the following ones:

- **Dublin Core**: Dublin Core (DC)[9] is a minimal metadata element set mainly used for the description of web resources. The DC vocabulary is available as an RDFS

---

[5] Schemapedia: www.schemapedia.com

[6] Watson: watson.kmi.open.ac.uk/WatsonWUI/

[7] Swoogle: swoogle.umbc.edu

[8] Linked Open Vocabularies: lov.okfn.org

[9] DCMI Metadata Terms: www.dublincore.org/documents/dcmi-terms/

ontology and contains classes and properties, such as agent, bibliographic resource, creator, title, publisher, description and so on. The DC vocabulary elements are widely used not only in the bibliographical domain but throughout the entire Linked Data landscape. The DC ontology is partitioned in two sets: simple DC, which contains 15 core properties, and qualified DC, which contains all the classes and the rest of the properties, which are specializations of those in simple DC.

- **Friend-Of-A-Friend (FOAF)**: The Friend-Of-A-Friend (FOAF)[10] vocabulary has been one of the first lightweight ontologies being developed since the first years of the Semantic Web. It mainly describes human social networks, including classes such as Person, Agent or Organization and properties denoting personal details and relationships with other persons and entities.
- **Simple Knowledge Organization System (SKOS)**: SKOS[11] is a fundamental vocabulary in the Linked Data ecosystem, containing, as its name suggests, terms for organizing knowledge in the form of taxonomies, thesauri and hierarchies of concepts. SKOS is a W3C recommendation since 2009 and its use is widespread in the library and information science domain. SKOS is defined as an OWL Full ontology and has its own semantics, i.e. its own set of entailment rules distinct from those of RDFS and OWL. SKOS defines *concept schemes* as sets of *concepts*, which can relate to each other through "broader/narrower than" or equivalence relationships.
- **Vocabulary of Interlinked Datasets (VoID)**: VoID is a W3C-proposed vocabulary that provides metadata on a given RDF dataset. With the use of VoID, a dataset owner or administrator can provide general metadata about that dataset (e.g. its name, description, creator), information on the ways that this dataset can be accessed (e.g. as a data dump or via a SPARQL endpoint), structural information on this dataset (e.g. the set of vocabularies that are used or the pattern of a typical URI), as well as information on the links that exist in this dataset (e.g. the number and target datasets of those links) (Alexander et al. 2009). VoID descriptions aim to aid users and machines to decide whether a dataset is suitable for their information needs and a particular task at hand.
- **Semantically-Interlinked Online Communities (SIOC)**: SIOC[12] is an OWL ontology that describes online communities, such as forums, blogs and mailing lists. Main classes describe concepts like forum, post, event, group and user, while properties denote attributes of those classes, such as the topic or the number of views of a post.
- **Good Relations**: Good Relations[13] is a vocabulary that describes online commercial offerings, covering issues that range from product and business descriptions to pricing and delivery methods. Good Relations is one of those vocabularies with the greatest impact in real-life applications, since it has been adopted by

---

[10] FOAF Vocabulary Specification: xmlns.com/foaf/spec/

[11] SKOS Reference www.w3.org/TR/skos-reference/

[12] SIOC: www.sioc-project.org

[13] GoodRelations: purl.org/goodrelations/

several online retailers and search engines. The latter are able to interpret product metadata expressed in the Good Relations vocabulary and offer results that are better tailored to the needs of end users.

Datasets can be classified to those that contain statements pertaining to a single domain and those that are essentially cross-domain, involving facts from several interrelated domains. Since the Linked Data cloud has grown to billion-triple levels, several "thematic neighborhoods" have begun to emerge in it. Among the domains that represent a significant percentage of knowledge in the LOD cloud, government, media, geography and life sciences are the most notable ones.

- **DBpedia**[14] is the RDF version of the Wikipedia knowledge base and perhaps the most popular RDF dataset in the LOD cloud, since there is a large number of incoming links from other datasets pointing to entities described in it. DBpedia is a cross-domain dataset that assigns a URI to every resource described by a Wikipedia article and produces structured information by mining information from Wikipedia infoboxes. This infobox information is encoded in an ontology, terms from which are used in DBpedia statements. DBpedia is also a multilingual dataset, since it transforms various language editions of Wikipedia. The DBpedia dataset is offered as data dump and through a SPARQL endpoint, while there is also a "live" version, which is updated whenever a Wikipedia page is updated.
- **Freebase**[15] is an openly-licensed structured dataset that is also available as an RDF graph. This cross-domain RDF dataset contains information of tens of millions of concepts, types and properties. Freebase is a dataset that can be edited by anyone, gathers information from several structured sources as well as free text and offers an API for developers who want to exploit this large knowledge base. Freebase is linked to DBpedia through both incoming and outgoing links.
- **GeoNames**[16] is an open geographical database containing millions of geographical places that can be edited by anyone. Geographical information usually provides the context of descriptions and facts and it proves to be omnipresent in the Linked Data cloud, leading to the existence of several incoming links from other datasets.
- **Lexvo**[17] is a linguistic dataset that provides identifiers for thousands of languages, URIs for terms in every language, as well as identifiers for language characters. Lexvo constitutes the central dataset for the so-called linguistic Linked Data Cloud.

Further insight in the various trends that LOD datasets follow (for example, the vocabularies being reused or the amount of self-describing metadata) can be found in the State of the LOD Cloud document.[18]

---

[14] DBpedia: www.dbpedia.org

[15] Freebase: www.freebase.com

[16] GeoNames: www.geonames.org

[17] Lexvo.org: www.lexvo.org

[18] State of the LOD Cloud: linkeddatacatalog.dws.informatik.uni-mannheim.de/state/

References

Alexander K, Cyganiak R, Hausenblas M et al (2009) Describing linked datasets—on the design and usage of voiD, the 'vocabulary of interlinked datasets'. In: Bizer C, Heath T, Berners-Lee T et al (eds) Linked Data on the Web (LDOW 2009). Proceedings of the WWW2009 workshop on Linked Data on the Web, Madrid, Spain, April 2009. CEUR workshop proceedings, vol 538

Arenas M, Bertails A, Prud'hommeaux E et al (2012) A direct mapping of relational data to RDF. World Wide Web Consortium. http://www.w3.org/TR/rdb-direct-mapping/. Accessed 24 Dec 2014

Brachman RJ, Schmolze JG (1985) An overview of the KL-ONE knowledge representation system. Cognit Sci 9:171–216

Brickley D, Guha RV (2014) RDF Schema 1.1. World Wide Web Consortium. http://www.w3.org/TR/rdf-schema/. Accessed 24 Dec 2014

Carothers G (2014) RDF 1.1 N-Quads: a line-based syntax for RDF datasets. World Wide Web Consortium. http://www.w3.org/TR/n-quads/. Accessed 24 Dec 2014

Carothers G, Seaborne A (2014) RDF 1.1 N-Triples: a line-based syntax for an RDF graph. World Wide Web Consortium. http://www.w3.org/TR/n-triples/. Accessed 24 Dec 2014

Carothers G, Seaborne A (2014) RDF 1.1 TriG: RDF Dataset Language. World Wide Web Consortium. http://www.w3.org/TR/trig/. Accessed 24 Dec 2014

Connolly D (2007) Gleaning Resource Descriptions from Dialects of Languages (GRDDL). World Wide Web Consortium. http://www.w3.org/TR/grddl/. Accessed 24 Dec 2014

Das S, Sundara S, Cyganiak R (2012) R2RML: RDB to RDF mapping language. World Wide Web Consortium. http://www.w3.org/TR/r2rml/. Accessed 24 Dec 2014

Gandon F, Schreiber G (2014) RDF 1.1 XML syntax. World Wide Web Consortium. http://www.w3.org/TR/rdf-syntax-grammar/. Accessed 24 Dec 2014

Harris S, Seaborne A (2013) SPARQL 1.1 query language. World Wide Web Consortium. http://www.w3.org/TR/sparql11-query/. Accessed 24 Dec 2014

Hendler J (2008) Web 3.0: Chicken farms on the Semantic Web. Computer 41(1):106–108

Herman I, Adida B, Sporny M, Birbeck M (2013) RDFa 1.1 Primer: Rich Structured Data Markup for Web Documents. World Wide Web Consortium. http://www.w3.org/TR/rdfa-primer/. Accessed 24 Dec 2014

Hitzler P, Krötzsch M, Parsia B et al (2012) OWL 2 Web Ontology Language Primer. World Wide Web Consortium. http://www.w3.org/TR/owl2-primer/. Accessed 24 Dec 2014

Horridge M, Drummond N, Goodwin J et al (2006) The Manchester OWL syntax. In: Cuenca Grau B, Hitzler P, Shankey C et al (eds) OWLED'06 OWL: experiences and directions 2006. OWLED'06 workshop, Athens, Georgia, USA, November 2006. CEUR workshop proceedings, vol 216

Horrocks I, Patel-Schneider P, van Harmelen F (2003) From SHIQ and RDF to OWL: the making of a Web Ontology Language. J Web Semantics 1(1):7–26

Kifer M, Boley H (2010) RIF overview. World Wide Web Consortium. http://www.w3.org/TR/rif-overview/. Accessed 24 Dec 2014

Knublauch H, Hendler JA, Idehen K (2011) SPIN—overview and motivation. World Wide Web Consortium. http://www.w3.org/Submission/spin-overview/. Accessed 24 Dec 2014

McGuinness D, van Harmelen F (2004) OWL Web Ontology Language overview. World Wide Web Consortium. http://www.w3.org/TR/owl-features/. Accessed 24 Dec 2014

Prud'hommeaux E, Carothers G (2014) RDF 1.1 Turtle: Terse RDF Triple Language. World Wide Web Consortium. http://www.w3.org/TR/turtle/. Accessed 24 Dec 2014

Scheppe K (2009) Protocol for Web Description Resources (POWDER): Primer. World Wide Web Consortium. http://www.w3.org/TR/powder-primer/. Accessed 24 Dec 2014

Schreiber G, Raimond Y (2014) RDF 1.1 Primer. World Wide Web Consortium. http://www.w3.org/TR/rdf11-primer/. Accessed 24 Dec 2014

Speicher S, Arwe J, Malhotra A (2014) Linked Data Platform 1.0. World Wide Web Consortium. http://www.w3.org/TR/ldp/. Accessed 24 Dec 2014

Sporny M, Kellogg G, Lanthaler M (2014) JSON-LD 1.0: a JSON-based serialization for Linked Data. World Wide Web Consortium. http://www.w3.org/TR/json-ld/. Accessed 24 Dec 2014

Chapter 3
Deploying Linked Open Data: Methodologies and Software Tools

3.1 Introduction

As introduced in the previous chapters, the ultimate goal of the Semantic Web is to bring order in the chaos of online information so as to enable the web user and future applications to effectively search and find accurate information. Among the major problems with today's Web is that mostly 'anyone can say anything about any topic', meaning that any individual is given the freedom to express any piece of information combined with information from any other source (Allemang and Hendler 2008). As a result, information that can be found on the Web cannot always be trusted.

In order to provide a solution to this problem, the Linked Open Data (LOD) approach is introduced, as it effectively materializes the Semantic Web vision. Linked data allows querying across different data sources. In the simplest sense, a focal point is provided for referencing (referring to) and de-referencing (retrieving data about) any given web resource. Having introduced in the previous chapter the main technologies involved in LOD creation, in this chapter, we further analyze the issue of publishing Linked Open Data, by presenting, discussing and looking into the details of the basic methodologies and tools that serve this purpose.

Before continuing, it is first necessary to recognize that not all data(sets) are suitable for online publishing. They have to demonstrate some or all of the following properties, from an application development perspective (Bizer et al. 2009):

- First of all, data has to be stand-alone, in the sense that it is strictly separated from any business logic, formatting or presentation processing.
- Data has to be adequately described, in order to allow for third parties to consume it. This is achieved by using well-known vocabularies to describe the dataset, or, in cases when the vocabulary is somewhat less popular, by providing de-referenceable URIs with vocabulary term definitions.
- The dataset is encouraged to be open, in the sense that it can contain RDF links to other datasets which can be discovered at run-time by following these links.

© Springer International Publishing Switzerland 2015
N. Konstantinou, D.-E. Spanos, *Materializing the Web of Linked Data*,
DOI 10.1007/978-3-319-16074-0_3

- HTTP being used as the data access mechanism and RDF as the structural data model greatly simplifies data access compared to Web APIs which may rely on heterogeneous data models and access interfaces.

Linked Data-driven applications can be grouped into four categories (Hausenblas 2009):

- *content reuse applications*, such as BBC's Music store that (re)uses metadata from DBpedia, and MusicBrainz[1]
- *semantic tagging and rating applications* such as Faviki[2] that uses unambiguous identifiers from DBpedia
- *integrated question-answering systems*, such as DBpedia mobile (Becker and Bizer 2008) able to indicate locations from the DBpedia dataset in the user's vicinity
- *event data management systems*, such as Virtuoso's (see Sect. 3.4.2.2) calendar module that can organize events, tasks, and notes.

Using the Linked Data approach, data webs are also expected to evolve in numerous domains, from biology (Zhao et al. 2009) to software engineering (Iqbal et al. 2009). However, the bulk of Linked Data processing, which is the creation of the semantically annotated data source, is not done online. Traditional applications have datasets stored and processed using other technologies (relational databases, spreadsheets, XML files, etc.) which must be transformed into LOD datasets in order to be published on the web. We continue our discussion in this chapter by discussing how to convert existing data into LOD, or how to generate LOD directly, by presenting tools, methodologies and technological challenges associated with the task.

This chapter is structured as follows: Sect. 3.2 discusses openness and respective approaches, Sect. 3.3 discusses content modeling, making a special reference to the identifier assignment problem. Section 3.4 provides a comprehensive list of the most popular software tools and libraries available for authoring, cleaning up, storing, processing and linking data.

3.2 The O in LOD: Open Data

How is Linked Data related to Open Data? First of all, it needs to be defined that these are two completely different concepts. Open data is data that is publicly accessible via internet, without any physical or virtual barriers to accessing them. Linked Data now, is data that allows relationships to be expressed among these data. RDF is ideal for representing Linked Data, which is something that contributes to the general misconception that LOD can only be published using RDF.

---

[1] MusicBrainz: www.musicbrainz.org

[2] Faviki: www.faviki.com

Having so far described how to *link* the data, we have to shift the focus to what does *opening* the data mean. A widely adopted definition of openness in relation to data and content is provided by www.opendefinition.org, and it is summed up in the following statement:

> "Open means anyone can freely access, use, modify, and share for any purpose (subject, at most, to requirements that preserve provenance and openness)".

Now, why should anyone open their data? Not a long time ago, data owners such as government agencies, were reluctant, if not opposed, to the idea of opening their content as available for public access. This was happening for various reasons. Especially in organizations whose core business was the maintenance and enrichment of data archives, the biggest reason for reluctance was the fear of becoming useless by giving away their core value. Experience has shown that in practice the opposite happens: Allowing access to one's content leverages its value because of a series of reasons:

- Third parties and interested audience can reuse the data by creating added-value services and products on top of the content
- Mistakes and inconsistencies can be discovered more easily when more people access to the content and can verify its freshness, completeness, accuracy, integrity and overall value.

In specific domains, data have to be open for strategic reasons, e.g. regarding Government Data, an important aspect is transparency: data that is collected by the public should be made available to the public.

The task of publishing Linked Open Data includes several steps. In order to maintain consistency, provide actual additional value and maximize reuse potential, several recommendations exist in the literature, including the following[3]:

- **Data should be kept simple**. In order to keep it simple, you could start small and fast, as it is not required that all data is opened at once. You could start by opening up just one dataset, or even a part of a larger dataset. It is certainly better to open up more datasets because experience and momentum may be gained, but opening datasets is also a risky choice, since it may lead to unnecessary spending of resources (e.g. not every dataset is useful).
- **Engage early and engage often**. As in any business domain, it is crucial that you know your audience (potential users), receive feedback and take it into account. This means that service providers must engage with end users and ensure that the next iteration of the service will be as relevant as it can be. In many cases, end users will not be direct consumers of the data; it is likely that intermediaries will come between data providers and end users. For instance, an

[3] The Open Knowledge Foundation: How to Open Data: okfn.org/opendata/how-to-open-data/

end user will not find use in an array of geographical coordinates but a company offering maps will. Thus, engaging with the intermediaries is also crucial, as they will reuse and repurpose the data.

- **Common fears and misunderstandings** should be dealt with in advance. Especially in large institutions, it is not granted that opening up data will be looked upon favorably, as it will entail a series of consequences and, respectively, opposition. Therefore, from an early stage, the most important fears and probable misconceptions should be identified, explained, and be dealt with.

It is fine to charge for access to the data via an API, as long as the data itself is provided in bulk for free. Therefore, data can be considered as open, while the API is considered as an added-value service on top of the data, which makes sense to charge fees for its usage (usage of the API, not of the data). This opens business opportunities in the data-value chain around open data.

Data openness is a different concept than data freshness. This means that the data that is opened does not necessarily have to be a real-time snapshot of the system data. Especially in cases when data is streamed through e.g. a sensor or a social network, it is considered ok to provide data that is consolidated into bulks asynchronously, e.g. every hour or every day. One approach can be to offer bulk access to the data dump and access through an API to the real-time data.

Next, it is not obligatory but it is important that some metadata be provided about the dataset itself. Namely:

1. Its provenance. Provenance is information about entities, activities and people involved in the creation of a dataset, a piece of software, a tangible object, a thing in general. This information is important as it can be used in order to assess the thing's quality, reliability, trustworthiness, etc. In order to allow a uniform manner to express provenance-specific information, the W3C provides two related recommendations: (a) The PROV Data Model,[4] expressed in OWL 2 by (b) the PROV ontology.[5]
2. Description about the dataset. DCAT[6] is a W3C recommendation describing an RDF vocabulary specifically designed to facilitate interoperability between data catalogs published on the Web.
3. Licensing. It is important to provide licensing information, for instance in the web page that offers the dataset, a short description regarding the terms of use of the dataset. For instance, if the Open Data Commons Attribution License is used, a small text of the following form needs to be present, stating:

> This {DATA(BASE)-NAME} is made available under the Open Data Commons Attribution License: http://opendatacommons.org/licenses/by/{version}.—See more at: http://opendatacommons.org/licenses/by/#sthash.9HadQzSW.dpuf

---

[4] The PROV Data Model: www.w3.org/TR/prov-dm/

[5] The PROV Ontology: www.w3.org/TR/prov-o/

[6] Data Catalog Vocabulary: www.w3.org/TR/vocab-dcat/

where the publisher replaces {DATA(BASE)-NAME} with the name of the data or database, and {version} with the license version.

3.2.1 Opening Data: Bulk Access vs. API

The next question that needs to be answered is whether the data should be provided via an API, such as a SPARQL endpoint (see Sect. 2.5) and/or a RESTful service, or to allow bulk access to it. According to (Open Knowledge 2012), offering bulk access to the data is a requirement but offering an API is not. It is, of course, great if an API is available. This is mainly because bulk access to the data, compared to providing an API, presents the following characteristics:

- Providing bulk access to the data can be cheaper than providing an API. Even an elementary API entails development and maintenance costs.
- Offering bulk access allows building an API on top of the offered data, but simply offering an API does not allow clients to retrieve the whole amount of data in order to process it.
- Bulk access guarantees full access to the data, an API does not.

An API is more suitable for large volumes of data. In case a client needs a small subset of the provided information, it is not necessary to download the whole dataset.

3.2.2 The 5-Star Deployment Scheme

As several factors are being altogether considered for LOD creation, it is obvious that one cannot state with a simple yes or no whether a dataset is open. For this reason, the following 5-star deployment scheme has been suggested for Linked Data,[7] according to the properties the resulting dataset presents:

★	Data is made available on the Web (whatever format) but with an open license to be Open Data
★★	Available as machine-readable structured data: e.g. an Excel spreadsheet instead of image scan of a table
★★★	As the 2-star approach, in a non-proprietary format: e.g. CSV instead of Excel
★★★★	All the above plus the use of open standards from W3C (RDF and SPARQL) to identify things, so that people can point at your stuff
★★★★★	All the above, plus: Links from the data to other people's data in order to provide context

---

[7] Linked Data: www.w3.org/DesignIssues/LinkedData.html

3.3   The D in LOD: Modeling Content

First, the content to be published has to follow certain norms, comply with a specific *model*. A model can be used as a mediator among multiple viewpoints, it can be used as an interfacing mechanism between humans or computers to understand each other, even offer useful analytics and predictions. In this section, we discuss the desired properties of a semantic data model. The model can be expressed in RDF(S), OWL, be custom or reuse existing vocabularies (see Sect. 2.8) etc. Therefore, among the first decisions that have to be taken when publishing a dataset as LOD is regarding the ontology that will serve as a model.

Now, designing vocabularies and ontologies is a domain that has existed long before even the emergence of the Web. As a result, widespread vocabularies exist in several domains, encoding the knowledge and experience that is accumulated in these domains. As a consequence, regardless of the domain of interest, it is highly probable that a vocabulary has already been created in order to describe the concepts involved (Archer et al. 2013). It is important to build on top of that work, without replicating it, for several reasons:

- **Increased interoperability**. Take for instance the example of encoding a date value. Adherence to standards can help content aggregators to parse and process the information without much extra effort per data source. Using, e.g. the DCMI Metadata Terms, a date field, e.g. dcterms:created can be encoded as "2014-11-07"^^xsd:date. A system burdened with e.g. the task of processing dates incoming from several sources, is very likely to support the standard date formats and less likely (as it is less efficient) to convert the formatting from each source to a uniform syntax. Therefore, exposing the data using standard vocabulary approaches, increases the interoperability potential.
- **Credibility**. It shows that the dataset that has been published has been taken care of, in the sense that it is well thought of, curated, and having performed a state-of-the-art survey prior to publishing the data.
- **Ease of use**. Reusing existing approaches is undoubtedly easier than rethinking and implementing again/replicating already existing solutions. Even more, in the case when the vocabularies are published by multidisciplinary consortia with potentially more spherical view on the domain than yours.

In conclusion, before adding any terms in our vocabulary, it is important to make sure that those terms do not already exist. In such case, the existing terms must be reused by reference. In situations when we need to be more specific, we can always create a subclass or a subproperty, based on the existing ones. New terms can always be generated, when the existing ones do not suffice.

Also, the complexity of the model has to be taken into account, based on the desired properties. This is in order to decide whether RDFS or one of the OWL profiles is needed. If the RDFS expressiveness does not cover the project needs, then the rich constructs of the OWL language should be used. In the case of OWL, then, one should be aware of its different flavors, and their different respective capabilities, covered in more detail in Sect. 2.4.2.

Overall, Semantic Web technologies can be used behind the scenes for system modeling, as they provide powerful means for system description (a concept hierarchy, a property hierarchy, a set of individuals, etc.). However, the strength of the Semantic Web is not restricted in concept description: model checking can be realized by the concurrent use of a reasoner, a practice that assures the creation of coherent, consistent models. Beyond semantic interoperability, the goal is to exploit the ontologies' inference support, the formally defined semantics, the support of rules, and logic programming in general (Kappel et al. 2006).

3.3.1 Assigning URIs to Entities

First, it is important to keep in mind that descriptions can be provided both for items/persons/ideas/things (in general) that exist outside of the Web and for things that exist online. For instance, a company can be described by its company page. Thus, in order to provide mere information on the company, two URIs will be needed: One for the company's website and one for the company itself. The company description may well be in an RDF document. Therefore, a strategy has to be devised in assigning URIs to entities; there are no deterministic approaches.

Among the most important challenges in publishing data is assigning URIs (identifiers) to it. Identifiers, as parts of the published URIs, are at the heart of how data becomes linked. In order to provide certain recommendations regarding their usage, the Open Data Institute (ODI) and Thomson Reuters provide a set of recommendations with respect to several identifier challenges for Open Data.[8] According to the recommendations, data publishers will have to deal with several challenges such as: dealing with ungrounded data, lack of reconciliation options, lack of identifier scheme documentation, proprietary identifier schemes, multiple identifiers for the same concepts/entities, inability to resolve identifiers, fragile identifiers, etc. These are indicative of the difficulties that occur when publishing LOD. The respective recommendations for these challenges can be found in the report.

However, despite the difficulties for assigning URIs, a correct implementation can be highly beneficial. Besides the benefits entailed by semantic annotation, the whole idea behind publishing data in the Web is making it discoverable, citable, since the value of the primary data increases, as the usage of its identifiers increases.

In order to assign URIs identifying things of interest, several *design patterns* exist in the literature; conventions for how URIs will be assigned to resources. As mentioned in (Dodds and Davis 2012), many of these patterns are also widely used in modern web frameworks and are in general applicable to web applications. These conventions can be combined and, of course, they can evolve and be extended over time. Their use is not restrictive, as each dataset has its own characteristics. However,

---

[8] ODI and Thomson Reuters: Creating Value with Identifiers in an Open Data World, October 2014, available online at theodi.org/guides/data-identifiers-white-paper

some upfront thought about identifiers is always beneficial. Identifier patterns that can be considered include the following:

- **Hierarchical URIs**: URIs assigned to a group of resources that form a natural hierarchy, e.g. `:collection/:item/:sub-collection/:item`.
- **Natural Keys**: URIs created from data that already has unique identifiers, e.g. using the their ISBN to identify books.
- **Literal Keys**: URIs created from existing, non-global identifiers. This can be addressed by using these identifiers to assign values to a property (e.g. the `dc:identifier` property from the Dublin Core vocabulary (see Sect. 2.8)) of the described resource.
- **Patterned URIs**: More predictable, human-readable URIs. For example `/books/12345`, where `/books` is the base part of the URI indicating "the collection of books", and the `12345` is an identifier for an individual book.
- **Proxy URIs**: URIs used in order to deal with the lack of standard identifiers for third-party resources. If for these resources, identifiers do exist, then these should be reused. Otherwise, locally minted Proxy URIs will have to be used.
- **Rebased URIs**: URIs constructed based on other URIs. For example, an application can use a regular expression to rewrite a URI `http://graph1.example.org/document/1` to `http://graph2.example.org/document/1`
- **Shared Keys**: URIs specifically designed to simplify the linking task between datasets. This is achieved by a creating *Patterned URIs* while applying the *Natural Keys* pattern, but preferring public, standard identifiers rather than internal, system-specific codes.
- **URL Slugs**: URIs created from arbitrary text or keywords, following a certain algorithm, such as lowercasing the text, removing special characters and replacing spaces with a dash. For instance, the creation of a URI for the name "Brad Pitt" could be `http://www.example.org/brad-pitt`.

Next, we deal with the technical approaches in URI assignment. Technically, the desired functionality for Semantic Web applications is the ability to retrieve the RDF description of things, whilst web browsers are directed to the (HTML) documents describing the same resource, as illustrated in Fig. 3.1. Now, in order to provide URIs for dataset entities, there are two broad categories, Hash URIs and 303 URIs (Ayers and Völkel 2008).

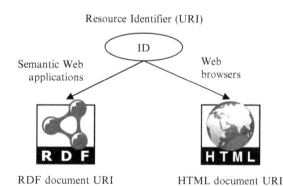

Fig. 3.1 Desired relationships between a resource and its representing documents (Ayers and Völkel 2008)

3.3.1.1   Hash URIs

In order to assign URIs to non-document resources, the URIs that are assigned, can contain a fragment, a special part that is separated from, the rest of the URI using the hash symbol ('#'). Suppose, for instance, we want to assign URIs to the descriptions of two companies, e.g. Alpha and Beta. Two valid URIs would be:

* http://www.example.org/info#alpha and
* http://www.example.org/info#beta

The RDF document that will contain descriptions about both companies will be http://www.example.org/info. This RDF document will be using the original URIs to uniquely identify the resources (i.e. companies Alpha, Beta and anything else).

The advantage of the Hash URIs is first that they reduce the number of HTTP round-trips performed by the client, which in turn, reduces access latency. Also, a family of URIs can share the same non-hash part, and the descriptions of companies Alpha and Beta can be retrieved with a single request to the RDF document. Two cases exist in following the Hash URIs approach: With and without content negotiation.

Hash URIs with content negotiation. Content negotiation can be employed to redirect from the about URI to either a RDF or an HTML representation, with the decision regarding which to return being based on client preferences and server configuration. Technically, the Content-Location header should be set to indicate if the hash URI refers to a part of the RDF document (info.rdf) or HTML document (info.html). Figure 3.2a shows the hash URI approach with content negotiation.

Hash URIs without content negotiation. This approach it not as technically challenging as the previous one, as it can be implemented by simply uploading static RDF files to a Web server, without any special server configuration. Therefore, this is a popular approach for quick-and-dirty RDF publication. However, the major problem of this approach is that clients interested in only one of the resources will be obliged to load (download) the whole RDF file. Figure 3.2b illustrates this simple approach.

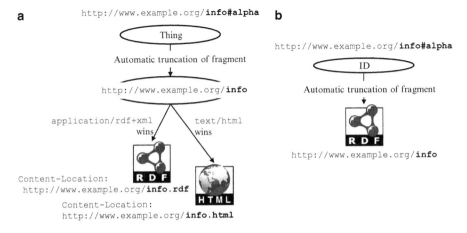

Fig. 3.2 The Hash URI generation approach. In (**a**) the Hash URI with content negotiation approach is illustrated while in (**b**) the Hash URI without content negotiation (Ayers and Völkel 2008)

3.3.1.2   303 URIs

The second approach on assigning URIs to resources relies on a special HTTP status code: `303 See Other`. This response is used in order to provide an indication that the requested resource is not a regular Web document.

The regular HTTP response (200) cannot be returned, because the requested resource does not have a suitable representation. However, we still can retrieve description about this resource. By doing so, we can distinguish between the real-world resource and its description (representation) on the Web.

The HTTP 303 is a redirect status code. Using it, the server can provide the location of a document that represents the resource. For instance, if in the example above, companies Alpha and Beta can be described using the following URIs:

http://www.example.org/id/alpha, and
http://www.example.org/id/beta

The server can be configured to answer requests to these URIs with a 303 (redirect) HTTP status code to a location containing an HTML, an RDF, or any alternative form, for example:

http://www.example.org/doc/alpha, and
http://www.example.org/doc/beta.

This setup allows to maintain bookmarkable, de-referenceable URIs for both the RDF and HTML views of the same resource. The 303 URIs approach, is a very flexible one, since the redirection target can be configured separately per resource. Also, there could be a document for each resource, or one (large) document with descriptions of all the resources.

Figure 3.3a illustrates the solution providing the generic document URI, while Fig. 3.3b shows the redirects for the 303 URI solution without the generic document URI.

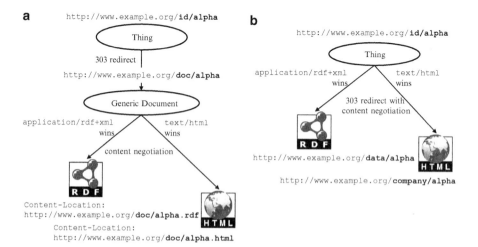

Fig. 3.3 The 303 URI generation approach. In (**a**) we see the approach in which URIs forward to one generic document, while in (**b**) the approach in which URIs forward to different documents (Ayers and Völkel 2008)

In Figure 3.3b, requests to the URI http://www.example.org/id/alpha will be redirected to an RDF (http://www.example.org/data/alpha) or an HTML (http://www.example.org/company/alpha) description of the resource. The RDF document would contain statements about the appropriate resource, using the original URI (http://www.example.org/id/alpha) to identify the described resource.

At all cases, the problem, however, resides in the induced latency caused by the redirects of the client. A client looking up a set of terms through 303 may use many HTTP requests, simply for learning that everything that could be loaded in the first request, is there and is ready to be downloaded.

In cases of large-scale datasets using the 303 approach, clients may be tempted to download the full data, using many requests. In these cases, SPARQL endpoints or comparable services should be provided in order to answer complex queries directly on the server, rather than having the clients downloading large data sets via HTTP. Finally, note that the 303 and Hash approaches are not mutually exclusive. Contrarily, combining them could be ideal in order to allow large datasets to be separated into multiple parts and have identifiers for non-document resources.

3.4 Software for Working with Linked Data

Working on small datasets is certainly a task that can be tackled by manually authoring an ontology. However, when publishing LOD means that the data has to be programmatically manipulated, many tools exist that facilitate the effort. This section lists the most prominent examples in the domain, starting with ontology authoring environments and a comment on the need for cleaning data. Next, several approaches are presented regarding software tools and libraries for working with Linked Data. Of course, it is often the case that clear lines cannot be drawn among software categories, as there are for instance graphical tools offering programmatic access, as well as there are software libraries offering application tools with a graphical interface, etc.

3.4.1 Ontology Authoring Environments

Ontology authoring is not a linear process, in the sense that it is not possible to line up the steps needed in order to complete its authoring. It can rather be considered as an iterative procedure, since the structure that will hold the core of the ontology concepts can be enriched with more specialized, peripheral concepts, thus further complicating concept relations. Therefore, the more the ontology authoring effort advances, the more complicated the ontology becomes. The desired result can be reached using various approaches, for instance, starting from the more general and continuing with the more specific concepts. Reversely, one could write down the more specific concepts and subsequently group them.

Overall, ontology authoring can uncover existing problems such as concept clarification, understanding of the domain in order to create its model, and probable reuse and connect to other ontologies, along with a viable, basic concept design.

Regarding ontology editors, since the purpose of creating an ontology is to describe a domain to people not familiar with it, it is very important to offer a graphical interface through which the user can interact, assuring syntactic validity of the ontology, without revealing to the user its textual representation, which is prohibitively obscure to the amateur user's eyes.

In addition to offering a graphical interface, an ontology editor also has to offer several facilities such as consistency checks. The necessary freedom that the author has to define concepts and their relations, has to be constrained in order not assure that the resulting ontology is semantically consistent. Also, revisions are bound to take place, most likely not by a person alone, since domain expertise is needed along with familiarity to technology and the respective tools.

This whole process is more difficult than it sounds, and this is why several tools have been built in order to facilitate it. However, despite the fact that there are several ontology editors, only a few are practically used, among them the tools presented next.

Protégé, maintained by the Stanford Center for Biomedical Informatics Research, is an open-source project which is to date among the most complete and capable solutions for ontology authoring and managing, and among the most long-lived projects in the domain. Protégé is among the most prominent ontology editors, featuring a rich set of plugins and capabilities. After several years of development, it offers now, apart from the traditional Protégé Desktop, the WebProtégé solution for collaborative viewing and editing.

The Desktop version of Protégé offers a customizable user interface for creating and editing ontologies, allowing multiple ontologies being developed in a single frame workspace. Several Protégé frames roughly correspond to OWL components, namely the Classes, Properties (that is three frames: Object Properties, Data Properties, and Annotation Properties), and Individuals frames. Additionally, a set of tools is provided for the visualization, querying, and refactoring of the ontology, as well as reasoning support. Protégé offers full support for OWL 2 and connection to Description Logic reasoners like HermiT (included) or Pellet.

Web Protégé is a much younger offspring of the Desktop version. It was designed for allowing collaborative viewing and editing, but this ability comes at the expense of a less feature-rich and currently more buggy user interface.[9]

Finally, Protégé can also be used for imposing SPARQL queries to an ontology, for testing purposes.

TopBraid Composer[10] is a commercial, multi-purpose semantic platform for the authoring and editing of RDF graphs and OWL ontologies, based on the popular Eclipse development platform. TopBraid Composer is being developed by TopQuadrant and comes in a number of editions, each with separate functionality.

---

[9] Both versions are available online at protege.stanford.edu

[10] Topbraid Platform Overview: www.topquadrant.com/technology/topbraid-platform-overview/

The Maestro Edition essentially offers an integrated semantic development environment that covers aspects of the Semantic Web stack, such as data modeling, querying and user interface building. The TopBraid Platform offers a series of adapters for the conversion of data that is expressed in various forms, such as XML, spreadsheets and relational databases, in RDF graphs. Specifically for the latter case, TopBraid Composer allows for the dynamic extraction of RDF graphs from it by running SPARQL queries on it. It also provides features for the persistence of RDF graphs in external triple stores. Among its most novel features, the ability to define rules and constraints and associate them with OWL classes stands out: these rules are expressed in SPIN, a technology that is briefly described in Sect. 2.7.

The free edition of TopBraid Composer merely offers a graphical interface for the definition of RDF graphs and OWL ontologies and the execution of SPARQL queries on them.

The NeOn Toolkit[11] is an open-source ontology authoring environment, also based on the Eclipse platform. NeOn Toolkit was mainly implemented in the course of the NeOn project, funded by the European Commission, and its main goal is the support for all tasks that comprise the ontology engineering life-cycle. NeOn Toolkit contains a number of plugins that form a multi-user collaborative ontology development environment, deal with the issue of ontology evolution through time, facilitate ontology annotation, allow for querying and reasoning with an ontology, and also support working with mappings between relational databases and ontologies. The last issue is deeply explored in Sect. 4.2, where the ODEMapster plugin of the NeOn Toolkit is also mentioned.

3.4.2 Platforms and Environments for Working with (RDF) Data

Data that is published as Linked Data is not always necessarily produced primarily in this form. As a result, Linked Data may be produced from files in hard drives, relational databases, legacy systems etc. There are, therefore, many options regarding how the information is to be transformed into RDF, according to its primary source, and subsequently managed.

In this section we present the software tools and libraries that are available in the Linked Data ecosystem for working with data: converting into Linked Data from other formats, cleaning up, storing, visualizing, and linking to other sources. Note that the problem of creating Linked Data from relational databases is a special case, discussed throughout Chap. 4, in which several approaches are presented, classified, analyzed and discussed.

---

[11] NeOn Toolkit: www.neon-toolkit.org

3.4.2.1 Cleaning-Up Data: OpenRefine

When working with datasets, it is often the case that data quality may be lower than expected in terms of homogeneity, completeness, validity, consistency, etc. In these cases, some prior processing has to take place before publishing the data. While **OpenRefine**[12] is not a Semantic Web tool per se, we provide a special reference to the tool in order to highlight the importance of the data quality aspect: it is not enough to provide data as Linked Data. Of equal if not of higher importance is the reassurance that the published data meets certain quality standards.

OpenRefine was created specifically to help working with messy data and can be used to improve data consistency. Its basic aim is to allow users to improve data quality. It is used in cases where the primary data source are files in tabular form (e.g. TSV, CSV, Excel spreadsheets) or structured as XML, JSON, or even RDF.

It allows importing data into the tool and also connect them to other sources, in order to consume it. Although it is a web application, its intended use is to run locally, on one's own computer in order to allow processing of sensitive data, thus increasing the privacy it offers. Initially developed as "Freebase Gridworks" by Metaweb Technologies, Inc., it was renamed Google Refine after Google acquired the company in 2010, and OpenRefine after its transition to a community-supported project since 2012.

OpenRefine can be used in order to clean data by removing duplicate records, separating multiple values that may reside in the same field, splitting multi-valued fields, identifying errors (isolated or systematic), and applying ad-hoc transformations through the use of regular expressions.

Currently, OpenRefine at its core does not support conversion from other sources to RDF, but this is achieved by RDF Refine,[13] an extension that incorporates the functionalities of exporting to RDF (Cyganiak et al. 2010) and RDF reconciliation (Maali et al. 2011). The RDF export functionality is based on describing the shape of the generated RDF graph through a template that uses values from the input spreadsheet. By means of a graphical interface, the end user can specify the structure of an RDF graph, i.e. the relationships that hold among resources, the form of the URI scheme that will be followed, and so on.

The reconciliation part of the extension offers a series of alternatives for discovering Linked Data entities that are related to the resources being referenced in the input spreadsheet. The reconciliation service of RDF Refine allows reconciliation of resources against an arbitrary SPARQL endpoint (with or without full-text search functionality), or via the Sindice API (see Sect. 3.4.2.3). In the first case, a predefined SPARQL query that contains the request label (i.e. the label of the resource to be reconciled) is sent to a specific SPARQL endpoint, while in the second case, a call to the Sindice API is directly made using the request label as input to the service.

---

[12] OpenRefine (formerly Google Refine): openrefine.org

[13] RDF Refine: refine.deri.ie

3.4.2.2 Software Tools for Storing and Processing Linked Data

The tools presented in this section are powerful storing and processing solutions. It should be noted, though, that their use is not restricted to these capabilities. Overall, the wealth of tools that has been developed to date, forms a mature ecosystem of technologies and solutions, covering practical problems such as programmatic access, storage, visualization, querying via SPARQL endpoints, etc.

Sesame[14] is essentially a framework for processing RDF data. Sesame is an open source, fully extensible and configurable with respect to storage mechanisms, RDF-oriented Java framework. It offers transaction support, RDF 1.1 support, storing and querying APIs and a RESTful HTTP interface supporting SPARQL. Sesame offers the Storage And Inference Layer (Sail) API, a low level system API for RDF stores and inferences, allowing for various types of storage and inference to be used.

OpenLink Virtuoso[15] is a universal server solution, offerings an RDF data management and a Linked Data server solution—in addition to its capabilities as a web application/web services/relational database/file server.

Virtuoso offers a free and a commercial edition. Virtuoso implements a quad (graph, subject, predicate, object) store. Graphs can be directly uploaded to Virtuoso, they can be transient (i.e. not materialized) RDF views on top of its relational database backend, or crawled from third party RDF sources, or non-RDF (using Sponger) sources. Also, Virtuoso offers several plugins, including one that enables full-text search and faceted browsing on the graphs in its quadstore.

The **Apache Marmotta**[16] project features a Linked Data Client: the LDClient library is a modular tool that can convert data from other formats into RDF, and it can be used by any Linked Data project, independent of the Apache Marmotta platform. It has the capability of retrieving resources from remote data sources and map their data to appropriate RDF structures. The project includes a number of different backends that provide access to various kinds of online resources such as Freebase (see Sect. 2.8), the Facebook graph API, RDFa-augmented HTML pages, and much more.

Callimachus[17] is an open source platform for the development of web applications, based entirely on RDF and Linked Data. Callimachus, which is also available in an enterprise closed-source edition, relies on XHTML and RDFa templates that are populated by the results of SPARQL queries executed against an RDF triple store. These populated templates constitute the human-readable web pages, which essentially make Callimachus a structured Content Management System (CMS) or, in other words, a Linked Data Management System. As such, Callimachus supports end users in creating, managing, navigating and visualizing Linked Data through appropriate front-end components.

---

[14] Sesame RDF processing framework: rdf4j.org

[15] Virtuoso server: virtuoso.openlinksw.com

[16] Apache Marmotta platform: marmotta.apache.org

[17] Callimachus platform: callimachusproject.org

Visualization software in Linked Data is important, as tools in this category should make no assumptions of technical expertise, with the main target the mainstream user, who may have limited or non-existent knowledge of Linked Data and the related ecosystem. Besides the visualization offered from the aforementioned tools themselves, there are also projects targeted specifically at visualizing the Linked Data domain. Among them, are **LodLive**,[18] a project that provides a navigator that uses RDF resources, relying on SPARQL endpoints, and **CubeViz**,[19] a faceted browser for statistical data, that relies on the RDF Data Cube vocabulary for representing statistical data in RDF. Also, generic graph visualization platforms such as **Gephi**[20] or **GraphViz**,[21] both open-source tools, can come to the rescue.

Apache Stanbol[22] is a semantic content management system which aims at extending traditional CMS's with semantic services. It features a set of reusable components, all of which offer their functionalities via a RESTful web service API that returns JSON, RDF and supports JSON-LD. Its features include ontology manipulation, content enhancement (semantic annotation), reasoning, and persistence of semantic information.

Stardog[23] is an RDF database, geared towards scalability, also offering capabilities such as reasoning, OWL 2 and SWRL support. Stardog is implemented in Java, it exposes APIs for Jena and Sesame, and offers bindings for its HTTP protocol in numerous languages, including Javascript, .Net, Ruby, Clojure, and Python. A commercial as well as a free community edition are offered.

3.4.2.3   The L in LOD: Tools for Linking and Aligning Linked Data

The emerging Web of Data concept is to materialise the Semantic Web vision, to advance from a Web of Documents to a Web of (Linked) Data (Bizer et al. 2008). Instead of the current experience where the users navigate among (HTML) pages, the main idea is to navigate among (RDF) data. In other words, just as the current web can be crawled by users (and search engines) through hyperlinks, the Web of Data can be crawled through RDF links. RDF links simply assert relationships between Web resources by forming triples according to the semantic web paradigm: (resource, property, resource) and the main difference from simple hyperlinks is that, unlike the latter, they possess some meaning.

The task of establishing these links among entities of the dataset under publication and resources from external datasets of the LOD cloud is a crucial one, since it enables smooth integration of the new dataset in the Web of data. This step is emphasized in several proposed Linked Data lifecycles (e.g. the LOD2 Linked Data

---

[18] LodLive project: en.lodlive.it

[19] CubeViz faceted browser: aksw.org/Projects/CubeViz.html

[20] Gephi: gephi.github.io

[21] GraphViz visualization software: www.graphviz.org

[22] Apache Stanbol: stanbol.apache.org

[23] Stardog graph database: stardog.com

lifecycle (Auer et al. 2012), since without it, all published RDF datasets would essentially be isolated islands in the "ocean" of Linked Data. The task of establishing links can be either a manual process (i.e. the knowledge engineer that publishes a dataset should identify the most appropriate datasets and external resources that correspond to the published information) or a semi-automatic one, which harnesses existing open-source tools developed for such a purpose. While the manual approach is more suitable for small and static datasets (Heath and Bizer 2011), in most cases the semi-automatic approach is the most appropriate, especially when considering large datasets.

Silk[24] is one of the most popular open-source tools for the discovery of links among RDF resources of different datasets (Volz et al. 2009). Silk is a link discovery framework, the main component of which is a link specification language that specifies the details and criteria of the matching process, i.e. the source and target RDF datasets, the conditions that resources should fulfill in order to be interlinked, the RDF predicate that will be used for the link, the pre-matching transformation functions and similarity metrics to be applied, and so on. Silk is available as both a command line tool and a graphical interface, with the latter enabling the user to define graphically the matching rules. There is also a cluster edition of Silk, especially useful for the discovery of links among large datasets, a task that requires a lot of processing time. Silk accepts both local and remote datasets, available through SPARQL.

LIMES[25] is a link discovery framework among RDF datasets, that is based on the mathematical theory of metric spaces (Ngonga Ngomo and Auer 2011). By leveraging this theory, LIMES is able to restrict the number of comparisons made between the instances of two datasets and therefore increases its time efficiency. The matching process of LIMES is also driven by a configuration file, which is more succinct than the one used in Silk. In a nutshell, the LIMES engine extracts instances and properties from both source and target datasets, stores them in a cache storage or memory and computes the actual matches, based on restrictions specified in the configuration file. LIMES offers a web interface that facilitates the authoring of the configuration file and can also be applied to local datasets.

Sindice[26] is service that can be used for the manual discovery of related identifiers. Sindice is an index of RDF datasets that have been crawled and/or extracted from semantically marked up Web pages and, as such, offers both free-text search and SPARQL query execution functionalities. Furthermore, Sindice exposes several APIs that enable the development of Linked Data applications that can exploit Sindice's crawled content.

DBpedia Spotlight (Mendes et al. 2011) is a tool that was designed for annotating mentions of DBpedia resources in text. As such, it provides an approach for linking information from unstructured sources to the LOD cloud through DBpedia.

---

[24] Silk framework: wifo5-03.informatik.uni-mannheim.de/bizer/silk/

[25] LIMES framework: aksw.org/Projects/LIMES.html

[26] Sindice: sindice.com

The tool's modular architecture, to date, comprises: A web application, a web service, an annotation and an indexing API in Java/Scala, and an evaluation module.

Sameas.org is an online service that retrieves related LOD entities from some of the most popular datasets. The service serves more than 150 million URIs, providing a REST interface that retrieves related URIs for a given input URI or label. The service in fact does what the name hints: accepts URIs as inputs from the user and returns URIs that may well be co-referent.

3.4.3 Software Libraries for Working with RDF

Jena[27] (Carroll et al. 2004) is an open-source Java API that allows building of Semantic Web and Linked Data applications. First developed and brought to maturity by the Hewlett Packard Labs, Jena is now a project developed and maintained by the Apache Software Foundation. After its first version, in 2000, Jena has evolved to comprise nowadays a set of various tools, offering in fact the most complete solution for programmatic access and management of Semantic Web resources, being today the most popular Java framework for ontology manipulation. Besides the core RDF API, Jena also offers

- ARQ, a SPARQL implementation
- TDB, a high-performance triple store solution based on a custom implementation of threaded B+ Trees. Its purpose is the efficient storage and querying of large volumes of graphs. Many of the persistent data structures in TDB use a custom implementation of B+ trees, that only provides for fixed length key and fixed length value, and in which there is no use of the value part in triple indexes. Because of the custom implementation, it performs faster than a relational database backend, allowing the implementation to scale much more, as it is demonstrated in (Konstantinou et al. 2014). Practically, TDB stores the dataset in a single directory in the computer's file system. A TDB instance mainly consists of:

  - a Node table that stores the representation of RDF terms
  - Triple and Quad indexes, where triples are used for the default graph and quads for the named graphs
  - The Prefixes table, that provides support for Jena's prefix mappings and serves mainly presentation and serialization of triples issues, and does not take part in query processing

- Fuseki, a SPARQL Server, part of the Jena framework, that offers access to the data over HTTP using RESTful services. Fuseki can be downloaded and extracted locally, and run as a server offering a SPARQL endpoint plus some REST commands to update the dataset.

---

[27] Jena framework: jena.apache.org

- OWL support. Jena offers coverage of the OWL language.
- Inference API. Using an internal rule engine, Jena can be used as a reasoner.

Apache Any23.[28] The Apache Anything To Triples (Any23) is a programming library, also offering a web service and a command line tool, with the ability to extract structured data in RDF from a variety of Web documents. A variety of input formats is supported, including RDF in various formats (RDF/XML, Turtle, Notation 3), RDFa, a variety of microformats (hCalendar, hCard, hListing, etc.), HTML 5 Microdata (such as schema.org), JSON-LD (Linked Data in JSON format), and support for content extraction following several vocabularies such as CSV, Dublin Core Terms, Description of a Career, Description Of A Project, Friend Of A Friend, GeoNames, ICAL, Open Graph Protocol, schema.org, VCard.

Redland[29] is a set of libraries, written in C, that provide support for programmatic management and storage of RDF graphs. Its main libraries include:

- Raptor, which provides a set of parsers and serializers that read and write RDF graphs expressed in a variety of formats and syntaxes, such as RDF/XML, Turtle, RDFa, N-Quads and many more.
- Rasqal, which supports the query process of RDF graphs, mainly through SPARQL, by including functions that manage various query syntaxes and execute queries.
- Redland RDF Library, which handles the RDF manipulation an storage issues.

Redland also allows the invocation of its functions through a variety of programming languages, namely Perl, PHP, Ruby and Python.

EasyRDF[30] is a PHP library for the consumption and production of RDF. The project offers built-in parsers and serializers for most RDF serializations. Also, it allows querying using SPARQL, and type mapping from RDF resources to PHP objects.

RDFLib[31] is a Python library for working with RDF. Numerous tools are relying on this library, offering numerous capabilities, including serialization formats, working with microformats, RDFa, the OWL 2 RL profile, using relational databases as a backend, wrappers for remote SPARQL endpoints, and several other plugins and extras, composing a rich feature software library.

The Ruby RDF Project[32] allows RDF to be processed using the Ruby language. Currently, the library offers numerous mechanisms for reading/writing in different RDF formats, microdata support, querying using SPARQL and using relational databases as a storage backend. There is also a storage adaptor for the Sesame triplestore, for the No-SQL database MongoDB and more.

---

[28] Apache Any23: any23.apache.org

[29] Redland RDF libraries: librdf.org

[30] EasyRDF library: www.easyrdf.org

[31] RDFLib: github.com/RDFLib

[32] Ruby RDF: ruby-rdf.github.io

dotNetRDF[33] is an open source library for RDF written in C#.Net, also offering ASP.NET integration. The library supports SPARQL, reasoning, as well as integration with third party triple stores, including Jena, Sesame, Stardog, Virtuoso, etc. Besides the developer API, the library also offers a suite of command line and graphical tools for working with RDF and SPARQL, including conversions between RDF formats, utilities for running a SPARQL server and submitting queries, managing any supported triple stores, etc.

References

Allemang D, Hendler J (2008) Semantic Web for the working ontologist: effective modeling in RDFS and OWL. In: Chapter 1: what is the Semantic Web? Morgan Kaufmann, San Francisco, pp. 1–15

Archer P, Loutas N, Goedertier S (2013) Cookbook for translating relational data models to RDF schemas. European Commission, The ISA Programme, Report

Auer S, Buhmann L, Dirschl C, Erling O, Hausenblas M, Isele R, Lehmann J, Martin M, Mendes PN, van Nuelen B, Stadler C, Tramp S, Williams H (2012) Managing the life-cycle of Linked Data with the LOD2 stack. In: 11th international Semantic Web conference (ISWC 2012)

Ayers D, Völkel M (2008) Cool URIs for the Semantic Web. World Wide Web Consortium. http://www.w3.org/TR/cooluris/. Accessed 1 Jan 2015

Becker C, Bizer C (2008) DBpedia mobile: a location-enabled linked data browser. In: The 1st international workshop on Linked Data on the web, Beijing, China

Bizer C, Heath T, Idehen K, Berners-Lee T (2008) Linked Data on the Web. In: Proceedings of the 17th international conference on World Wide Web (WWW'08). ACM, New York, pp. 1265–1266

Bizer C, Heath T, Berners-Lee T (2009) Linked Data—the story so far. Int J Semant Web Inf Syst 5(3):1–22

Carroll J, Dickinson I, Dollin C et al (2004) Jena: implementing the Semantic Web recommendations. In: Proceedings of the 13th international World Wide Web conference on alternate track papers & posters, New York City, April 2004. ACM, New York, pp. 74–83

Cyganiak R, Maali F, Peristeras V (2010) Self-service Linked Government Data with DCAT and Gridworks. In: Proceedings of the 6th international conference on semantic systems, I-SEMANTICS'10. ACM, New York, pp. 37:1–37:3

Dodds L, Davis I (2012) Linked Data patterns: a pattern catalogue for modelling, publishing, and consuming Linked Data. http://patterns.dataincubator.org. Accessed 30 Dec 2014

Hausenblas M (2009) Linked data applications—the genesis and the challenges of using linked data on the web. Technical report, DERI Galway. http://linkeddata.deri.ie/sites/linkeddata.deri.ie/files/lod-app-tr-2009-07-26_0.pdf. Accessed 30 Dec 2014

Heath T, Bizer C (2011) Linked Data: evolving the web into a global data space. Morgan and Claypool Publishers, Seattle

Iqbal A, Ureche O, Hausenblas M, Tummarello G (2009) LD2SD: linked data driven software development. In: Proceedings of the 21st international conference on software engineering and knowledge engineering (SEKE'09). pp. 240–245

Kappel G, Kapsammer E, Kargl H, Kramler G, Reiter T, Retschitzegger W, Schwinger W, Wimmer M (2006) Lifting metamodels to ontologies: a step to the semantic integration of modeling languages. In: ACM/IEEE 9th international conference on model driven engineering languages and systems (MoDELS'06). Springer, Berlin, pp. 528–542

---

[33] dotNetRDF: www.dotnetrdf.org

Konstantinou N, Kouis D, Mitrou N (2014) Incremental export of relational database contents into RDF graphs. In: 4th international conference on web intelligence, mining and semantics (WIMS'14), Thessaloniki, Greece, June 2014. ACM, New York

Maali F, Cyganiak R, Peristeras V (2011) Re-using cool URIs: entity reconciliation against LOD hubs. In: Proceedings of the 4th Linked Data on the web workshop (LDOW 2011)

Mendes P, Jakob M, García-Silva A, Bizer C (2011) DBpedia spotlight: shedding light on the web of documents. In: Proceedings of the 7th international conference on semantic systems (I-Semantics), Graz, Austria

Ngonga Ngomo A-C, Auer S (2011) LIMES—a time-efficient approach for large-scale link discovery on the web of data. In: Proceedings of the 22nd international joint conference on artificial intelligence (IJCAI'11), pp. 2312–2317

Open Knowledge Foundation (2012) The Open Data handbook. http://opendatahandbook.org/pdf/OpenDataHandbook.pdf. Accessed 30 Dec 2014

Volz J, Bizer C, Gaedke M, Kobilarov G (2009) Discovering and maintaining links on the web of data. In: Bernstein A, Karger DR, Heath T, Feigenbaum L, Maynard D, Motta E, Thirunarayan K (eds) The Semantic Web—ISWC 2009: 8th international Semantic Web conference. Lecture notes in computer science, vol 5823. Springer, Berlin, pp. 650–665

Zhao J, Miles A, Klyne G, Shotton D (2009) Linked data and provenance in biological data webs. Brief Bioinform 10(2):139–152

Chapter 4
Creating Linked Data from Relational Databases

4.1 Introduction

The issue of collaboration among relational databases and Semantic Web standards has been extensively investigated for over a decade now. The problem of bridging the gap between relational and logic models is not just a theoretical exercise but has significant practical value. The motivation and benefits that nurtured this research direction are multifold and range from the need to bootstrap the Semantic Web with a sufficiently large mass of data to easier database integration, ontology-based data access and semantic annotation of dynamic Web pages (Spanos et al. 2012). In the rich related literature, the investigation of the similarities and differences among relational databases and Semantic Web knowledge models is often mentioned as the problem of *database-to-ontology mapping* (Konstantinou et al. 2008). This term is often misleading, grouping under the same name several distinct problems, each with its own motivation and goal. In this Chapter, we attempt to bring some order in the related literature by classifying proposed approaches to clearly distinguished problem classes.

The Chapter is structured as follows: Section 4.2 provides a discussion about the motivation behind the problem and the benefits that arise from the implementation of solutions. Next, in Sect. 4.3, a classification of the mapping approaches is presented, while in Sect. 4.4 the problem of creating and populating a new ontology from a relational database is discussed. Section 4.5 presents a complete example in the case of scholarly information, while Sect. 4.6 concludes the Chapter with a discussion on the future outlook of the problem of extracting RDF graphs from relational databases.

© Springer International Publishing Switzerland 2015
N. Konstantinou, D.-E. Spanos, *Materializing the Web of Linked Data*,
DOI 10.1007/978-3-319-16074-0_4

4.2 Motivation-Benefits

In this Section, we gather all motivations and benefits that are related with the database-to-ontology mapping problem, in order to better understand the distinct goals and challenges of the various aspects of this problem.

Semantic annotation of dynamic web pages. An aspect of the Semantic Web vision is the emergence of a Web of Data, from the current Web of Documents. A straightforward way to achieve this would be to annotate HTML pages, and add semantic information on top of their existing content, which currently defines how data is presented and is only suitable for human consumption. HTML pages can be semantically annotated with terms from ontologies, making their content suitable for processing by software agents and web services. Such annotations have been facilitated considerably since the proposal of the RDFa recommendation that embeds in XHTML tags references to ontology terms. However, this scenario does not work quite well for dynamic web pages that retrieve their content directly from underlying databases: this is the case for CMS's, forums, wikis and other Web 2.0 sites (Auer et al. 2009). Dynamic web pages represent the biggest part of the World Wide Web, forming the so called *Deep Web* (Geller et al. 2008), which is not accessible to search engines and software agents, since these pages are generated in response to an HTTP request.

It has been argued that, due to the infeasibility of manual annotation of every single dynamic page, a possible solution would be to directly "annotate" the underlying database schema, insofar as the web page owner is willing to reveal the structure of his database. This "annotation" is simply a set of correspondences between the elements of the database schema and an already existing ontology that fits the domain of the dynamic page content (Volz et al. 2004). Once such mappings are defined, it would be fairly trivial to generate dynamic semantically annotated pages with the embedded content annotations derived in an automatic fashion.

Heterogeneous database integration. The resolution of heterogeneity is one of the most popular, longstanding issues in the database research field, which remains, to a large degree, unresolved. Heterogeneity occurs between two or more database systems when they use different software infrastructure, follow different syntactic conventions and representation models, or when they interpret differently the same or similar data (Sheth and Larson 1990). Resolution of the above forms of heterogeneity allows multiple databases to be integrated and their contents to be uniformly queried. In typical database integration architectures, one or more conceptual models are used to describe the contents of every source database, queries are posed against a global conceptual schema and, for each source database, a wrapper is responsible to reformulate the query and retrieve the appropriate data. Ontology-based integration employs ontologies instead of conceptual schemas and therefore, correspondences between source databases and one or more ontologies have to be defined (Wache et al. 2001). Such correspondences consist of mapping formulas that can be viewed under the LAV and GAV paradigms, introduced in Sect. 1.2.2, where the target schema is in fact an ontology.

Therefore, they express the terms of a source database as a conjunctive query against the ontology (LAV mapping), express ontology terms as a conjunctive query against the source database (GAV mapping), or state an equivalence of two queries against both the source database and the ontology (Global–local-As-View or GLAV mapping). The discovery and representation of mappings between relational database schemas and ontologies constitute an integral part of a heterogeneous database integration scenario (An et al. 2006).

Ontology-based data access. Much like in a database integration architecture, ontology-based data access (OBDA) assumes that an ontology is linked to a source database, thus acting as an intermediate layer between the user and the stored data. The objective of an OBDA system is to offer high-level services to the end user of an information system who does not need to be aware of the obscure storage details of the underlying data source (Poggi et al. 2008). The ontology provides an abstraction of the database contents, allowing users to formulate queries in terms of a high-level description of a domain of interest. In some way, an OBDA engine resembles a wrapper in an information integration scenario in that it hides the data source-specific details from the upper levels by transforming queries against a conceptual schema to queries against the local data source. This query rewriting is performed by the OBDA engine, taking into account mappings between a database and a relevant ontology describing the domain of interest. The main advantage of an OBDA architecture is the fact that semantic queries are posed directly against a database, without the need to replicate its entire contents in RDF.

Apart from OBDA applications, a database-to-ontology mapping can be useful for *semantic rewriting of SQL queries*, where the output is a reformulated SQL query better capturing the intention of the user (Ben Necib and Freytag 2005). This rewriting is performed by substitution of terms used in the original SQL query with synonyms and related terms from the ontology. Another notable related application is the ability to query relational data using as context external ontologies (Das and Srinivasan 2009). This feature has been implemented in some database management systems, such as the Oracle and OpenLink Virtuoso DBMSs, allowing SQL queries to contain conditions expressed in terms of an ontology.

Mass generation of Semantic Web data. It has been argued that one of the reasons delaying the Semantic Web realization is the lack of successful tools and applications that will showcase the advantages of Semantic Web (SW) technologies (Konstantinou et al. 2010). The success of such tools, though, is directly correlated to the availability of a sufficiently large quantity of SW data, leading to a "chicken-and egg problem" (Hendler 2008), where cause and effect form a vicious circle. Since relational databases are one of the most popular storage media holding the majority of data on the World Wide Web, a solution for the generation of a critical mass of SW data would be the, preferably automatic, extraction of relational databases' contents in RDF. This would create a significant pool of SW data, that would alleviate the inhibitions of software developers and tool manufacturers and, in turn, an increased production of SW applications would be anticipated. The term database-to-ontology mapping has been used in the literature to describe such transformations as well.

Ontology learning. The process of manually developing from scratch an ontology is difficult, time-consuming and error-prone. Several semi-automatic ontology learning methods have been proposed, extracting knowledge from free and semi-structured text documents, vocabularies and thesauri, domain experts and other sources (Gómez-Pérez et al. 2003). Relational databases are structured information sources and, in case their schema has been modeled following standard practices (Elmasri and Navathe 2010) (i.e., based on the design of a conceptual model, such as UML or the Extended Entity Relationship Model), they constitute significant and reliable sources of domain knowledge. This is true especially for business environments, where enterprise databases are frequently maintained and contain timely data (Zhao and Chang 2007). Therefore, rich ontologies can be extracted from relational databases by gathering information from their schemas, contents, queries and stored procedures, as long as a domain expert supervises the learning process and enriches the final outcome. Ontology learning is a common motivation driving database-to-ontology mapping when there is not an existing ontology for a particular domain of interest, a situation that frequently arose not so many years ago. Nevertheless, as years pass by, ontology learning techniques are mainly used to create a wrapping ontology for a source relational database in an ontology-based data access (Sequeda and Miranker 2013) or database integration (Buccella et al. 2004) context.

Definition of the intended meaning of a relational schema. As previously mentioned, standard database design practices begin with the design of a conceptual model, which is then transformed, in a step known as logical design, to the desired relational model. However, the initial conceptual model is often not kept alongside the implemented relational database schema and subsequent changes to the latter are not propagated back to the former, while most of the times these changes are not even documented at all. Usually, this process leads to databases that have lost the original intention of their designer and are very hard to be extended or re-engineered to another logical model (e.g., an object-oriented one). Establishing correspondences between a relational database and an ontology grounds the original meaning of the former in terms of an expressive conceptual model, which is crucial not only for database maintenance but also for the integration with other data sources (Chebotko et al. 2009), and for the discovery of mappings between two or more database schemas (An et al. 2007). In the latter case, the mappings between the database and the ontology are used as an intermediate step and a reference point for the construction of inter-database schema mappings.

Integration of database content with other data sources. Transforming relational databases into a universal description model, as RDF aspires to be, enables seamless integration of their contents with information already represented in RDF. This information can originate from both structured and unstructured data sources that have exported their contents in RDF, thus overcoming possible syntactic disparities among them. The Linked Data paradigm (Heath and Bizer 2011), which encourages RDF publishers to reuse popular vocabularies in order to define links between their dataset and other published datasets and reuse identifiers that describe the same real-world entity, further facilitates global data source integration, regardless of the data source nature. Given the uptake of the Linked Data movement during the last few years, which has resulted in the publication of voluminous RDF content (in the order of billion statements) from several domains of interest, the anticipated bene-fits of the integration of this content with data currently residing in relational data-bases as well as the number of potential applications harnessing it are endless.

4.3   A Classification of Mapping Approaches

In this Section, we give a classification, which will help us categorize and analyze in an orderly manner approaches referring to the generic database-to-ontology map-ping problem. Furthermore, we introduce the descriptive parameters to be used for the presentation of each approach.

Classification schemes and descriptive parameters for database-to-ontology mapping methods have already been proposed in related work. The distinction between *classification criteria* and *descriptive measures* is often not clear. Measures that can act as classification criteria should have a finite number of values and ide-ally, should separate approaches in non-overlapping sets. Such restrictions are not necessary for descriptive features, which can sometimes be qualitative by nature instead of quantifiable measures. Table 4.1 summarizes classification criteria and descriptive parameters proposed in the related literature, despite the fact that some of the works mentioned have a somewhat different (either narrower or broader) scope than ours. As it can be seen from Table 4.1, there is a certain level of consensus among classification efforts, regarding the most prominent measures characterizing database-to-ontology mapping approaches.

The taxonomy we propose and which we are going to follow in the course of our analysis is based on some of the criteria mentioned in Table 4.1. The selection of these criteria is made so as to create a *total* classification of all relevant solutions in *mutually disjoint* classes. In other words, we partition the database-to-ontology mapping problem space in distinct categories containing uniform approaches. The few exceptions we come across are customizable software tools that incorporate multiple workflows, with each one falling under a different category.

We categorize solutions to the database-to-ontology mapping problem to the classes shown in Fig. 4.1. Next to each class, we append the applicable motivations and benefits, as mentioned in Sect. 4.2. Association of a specific benefit to a given class denotes that there is at least one work in this class mentioning this benefit.

The first major division of approaches is based on whether an existing ontology is *required* for the application of the approach. Therefore, the first classification criterion we employ is the **existence of an ontology** as a requirement for the map-ping process, distinguishing among methods that establish mappings between a given relational database and a given existing ontology (presented in Sect. 4.4.2) and methods that create a new ontology from a given relational database (presented

Table 4.1 Classification criteria and descriptive parameters identified in database-to-ontology mapping literature

Work	Classification criteria	Values	Descriptive parameters
Auer et al. (2009)	a. Automation in the creation of mapping	a. Automatic/Semi-automatic/ Manual	Mapping representation language
	b. Source of semantics considered	b. Existing domain ontologies/ Database/Database and User	
	c. Access paradigm	c. Extract-Transform-Load (ETL)/SPARQL/Linked Data	
	d. Domain reliance	d. General/Dependent	
Barrasa-Rodriguez and Gómez-Pérez (2006)	a. Existence of ontology	a. Yes (ontology reuse)/No (created ad-hoc)	–
	b. Architecture	b. Wrapper/Generic engine and declarative definition	
	c. Mapping exploitation	c. Massive upgrade (batch)/ Query driven (on demand)	
Ghawi and Cullot (2007)	a. Existence of ontology	a. Yes/No	Automation in the instance export process
	b. Complexity of mapping definition	b. Complex/Direct	
	c. Ontology population process	c. Massive dump/Query driven	
	d. Automation in the creation of mapping	d. Automatic/Semi-automatic/ Manual	
Hellmann et al. (2011)	–	–	Data source, Data exposition,
			Data synchronization
			Mapping language, Vocabulary reuse, Mapping automation
			Requirement of domain ontology, Existence of GUI
Konstantinou et al. (2008)	a. Existence of ontology	a. Yes/No	Ontology language
	b. Automation in the creation of mapping	b. Automatic/Semi-automatic/ Manual	RDBMS supported, Semantic query language
	c. Ontology development	c. Structure driven/Semantics driven	Database components mapped, Availability of consistency checks, User interaction

(continued)

Table 4.1 (continued)

Work	Classification criteria	Values	Descriptive parameters
Sahoo et al. (2009)	Same as in Auer et al. (2009) with the addition of:		
	a. Query implementation	a. SPARQL/SPARQL → SQL	Mapping accessibility, Application domain
	b. Data integration	b. Yes/No	
Sequeda et al. (2009)	–	–	Correlation of primary and foreign keys, OWL and RDFS elements mapped
Zhao and Chang (2007)	b. Database schema analysis	a. Yes/No	Purpose, Input, Output
			Correlation analysis of database schema elements
			Consideration of database
			Instance, application source
			Code and other sources

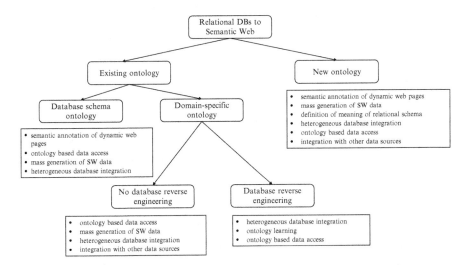

Fig. 4.1 Classification of database-to-ontology mapping approaches (Spanos et al. 2012)

in Sect. 4.4.1). In the former case, the ontology to be mapped to the database should model a domain that is compatible with the domain of the database contents in order for someone to be able to define meaningful mappings. This is the reason why the ontology is usually selected by a human user who is aware of the nature of data contained in the database. In the case of methods creating a new ontology, the existence of a domain ontology is not a prerequisite, including use cases where an ontology for the domain covered by the database is not available yet or even where the human user is not familiar with the domain of the database and relies on the mapping process to discover the semantics of the database contents.

The class of methods that create a new ontology is further subdivided in two subclasses, depending on the **domain of the generated ontology**. On the one hand, there are approaches that develop an ontology that has as domain the relational model itself. The generated ontology consists of concepts and relationships that reflect the constructs of the relational model, essentially mirroring the structure of the input relational database. We call such an ontology a "*database schema ontology*". Since the generated ontology does not implicate any other knowledge domain, these approaches are mainly automatic, relying on the input database schema alone. On the other hand, there are methods that generate domain-specific ontologies (presented in Sect. 4.4.1), depending on the domain described by the contents of the input database.

The last classification criterion we consider is the **application of database reverse engineering** techniques in the process of creating a new domain-specific ontology. Approaches that apply reverse engineering try to recover the initial conceptual schema from the relational schema and translate it to an ontology expressed in a target language. On the other hand, there are methods that, instead of reverse engineering techniques, apply few basic translation rules from the relational to the RDF model (detailed in Sect. 4.4.1) and/or rely on the human expert for the definition of complex mappings and the enrichment of the generated ontology model.

By inspection of Fig. 4.1, we can also see that, with a few exceptions, there is not significant correlation between the taxonomy classes and the motivations and benefits presented in Sect. 4.2. This happens because, as we argued above, this taxonomy categorizes approaches based on the *nature* of the mapping and the *techniques applied* to establish the mapping. On the contrary, most benefits state the *applications* of the already established mappings (e.g., database integration, web page annotation), which do not depend on the mapping process details. Notable exceptions to the above observation are the ontology learning and definition of semantics of database contents motivations. The former, by definition, calls for approaches that produce an ontology by analyzing a relational database, while the latter calls for approaches that ground the meaning of a relational database to clearly defined concepts and relationships of an existing ontology.

Given the taxonomy of Fig. 4.1 and the three classification criteria employed in it, we choose among the rest of the features and criteria mentioned in Table 4.1 the most significant ones as descriptive parameters for the approaches reviewed in this

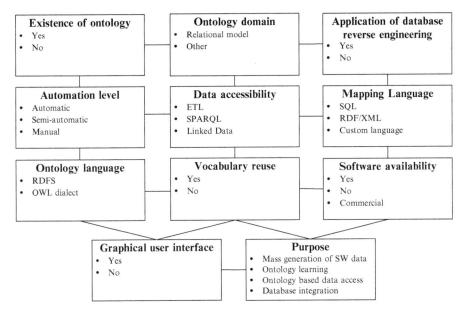

Fig. 4.2 Classification criteria and descriptive parameters for database-to-ontology mapping approaches (Spanos et al. 2012)

paper. The classification criteria and descriptive parameters adopted in this paper are shown in Fig. 4.2 and elaborated in the following:

Level of automation. This variable denotes the extent of user involvement in the process. Possible values for it are *automatic*, *semi-automatic* and *manual*. Automatic methods do not involve the user in the process, while semi-automatic ones require some input from the user. This input may be necessary for the operation of the process, constituting an integral part of the algorithm, or it may be optional feedback for the validation and enrichment of the result of the mapping process. In manual approaches, the mapping is defined in its entirety by the human user without any feedback or suggestions from the application. In some cases, automation level is a characteristic that is common among approaches within the same taxonomy class. For instance, methods that create a new database schema ontology are purely automatic, while the majority of approaches that map a relational database to an existing ontology are manual, given the complexity of the discovery of correspondences between the two.

Data accessibility. Data accessibility describes how the mapping result is accessed. This variable is also known in the literature as access paradigm, mapping implementation or data exposition. Possible values for this variable are ETL (Extract, Transform, Load), SPARQL or another ontology query language and Linked Data. ETL means that the result of the mapping process, be it an entire new ontology or a group of RDF statements referencing existing ontologies, is generated and stored as a whole in an external storage medium (often a triple store but it can be an external file as well). In this case, the mapping result is said to be *materialized*. Other terms used for ETL is batch transformation or massive dump. An alternative way of accessing the result of the mapping process is through SPARQL or some alternative semantic query language. In such a case, only a part (i.e., the answer to the query) of the entire mapping result is accessed and no additional storage medium is required since the database contents are not replicated in a materialized group of RDF statements. The semantic query is rewritten to an SQL query, which is executed and its results are transformed back as a response to the semantic query. This access mode is also known as query driven or on demand and characterizes ontology-based data access approaches. Finally, the Linked Data value means that the result of the mapping is published according to the Linked Data principles: all URIs use the HTTP scheme and, when dereferenced, provide useful information for the resource they identify.

The data accessibility feature is strongly related to *data synchronization*, according to which, methods are categorized in static and dynamic ones, depending on whether the mapping is executed only once or on every incoming user query, respectively. We believe that the inclusion of data synchronization as an additional descriptive feature would be redundant, since the value of data accessibility uniquely determines whether a method maintains static or dynamic mappings. Hence, an ETL value signifies static mappings, while SPARQL and Linked Data values denote dynamic mappings.

Mapping language. This feature refers to the language in which the mapping is represented. The values for this feature vary considerably, given that, until recently, no standard representation language existed for this purpose and most approaches used proprietary formats. W3C recognized this gap and proposed R2RML (see Sect. 2.6) as a language for expressing relational database-to-ontology mappings. The mapping language feature is applicable only to methods that need to reuse the mapping. A counterexample is the class of ontology learning methods that create a new domain ontology. Methods of this class usually do not represent the mapping in any form, since their intention is simply the generation of a new domain ontology, having no interest in storing correspondences with database schema elements.

Ontology language. This parameter states the language on which the involved ontology is expressed. Depending on the kind of approach, it may refer to the language of the ontology generated by the approach or to the language of the existing ontology required. Since the majority of methods reviewed implicate Semantic Web ontologies, the values for this feature range from RDFS to all flavours of OWL. When the expressiveness of the OWL language used is not clear from the description of the method, no specific profile or species is mentioned.

Vocabulary reuse. This parameter states whether a relational database can be mapped to more than one existing ontologies or not. This is mainly true for manual approaches where the human user is usually free to reuse terms from existing ontologies at will. Reusing terms from popular vocabularies is one of the prerequisites for Semantic Web success and hence, is an important aspect for database-to-ontology mapping approaches. Of course, methods allowing for vocabulary reuse do not *enforce* the user to do so, thus this variable should not be mistaken with the existence of ontology *requirement* that was used as a classification criterion in Fig. 4.1. Once again, values of this parameter may be common among all approaches within a certain class, e.g., methods generating a new database schema ontology do not directly allow for vocabulary reuse, since the terms used are specified a priori by the method and are not subject to change.

Software availability. This variable denotes whether an approach is accompanied with freely accessible software, thus separating purely theoretical methods or unreleased research prototypes from approaches providing practical solutions to the database-to-ontology mapping problem. Therefore, this feature serves (together with the *purpose* feature) as an index of the available software that should be considered from a potential user seeking support for a particular problem. Commercial software, which is not offering freely its full functionality to the general public, is indicated appropriately.

Graphical user interface. This feature applies to approaches accompanied with an accessible software implementation (with the exception of unavailable research prototypes that have been stated to have a user interface) and shows whether the user can interact with the system via a graphical user interface. This parameter is significant especially for casual users or even database experts who may not be familiar with Semantic Web technologies but still want to bring their database into Semantic Web. A graphical interface that guides the end user through the various steps of the mapping process and provides suggestions is considered as essential for inexperienced users.

Purpose. This is the main motivation stated by the authors of each approach. This is not to say that the motivation stated is the only applicable and that the approach cannot be used in some other context. Usually, all the benefits and motivations showcased in Fig. 4.1 apply to all tools and approaches of the same category.

We note once again that any of the above measures could have been considered as a criterion for the definition of an alternative taxonomy. However, we argue that the criteria selected lead to a partition of the entire problem space in clusters of approaches that present increased intra-class similarity. For example, some works (Auer et al. 2009), (Barrasa-Rodriguez and Gómez-Pérez 2006), (Ghawi and Cullot 2007), (Sahoo et al. 2009) categorize methods based on the data accessibility criterion to static and dynamic ones. This segregation is not complete though, as there are methods that restrict themselves in finding correspondences between a relational database and an ontology without migrating data to the ontology nor offering access to the database through the ontology. This was the main reason behind the omission of this distinction in the classification of Fig. 4.1.

In this Chapter, we consider only approaches that can be exploited for the publication of Linked Data from relational database instances. The majority of such approaches belongs to the "creation of domain-specific ontology" class, while there is also a number of "mapping discovery" methods that can also export database contents as an RDF graph, based on the discovered mappings. Approaches that create a "database schema ontology" are not immediately usable for publishing Linked Data, since the generated ontology essentially mirrors the relational structure and needs to be aligned with appropriate domain vocabularies. We also limit our attention to approaches that point to working implementations and can thus serve as practical solutions when extracting RDF graphs from relational databases.

4.4 Creating Ontology and Triples from a Relational Database

In this Section, we deal with the issue of generating a new ontology from a relational database and populating the former with RDF data that are extracted from the latter. Schematically, this process is roughly shown on Fig. 4.3. The mapping engine interfaces with the database to extract the necessary information and, according to manually defined mappings or internal heuristic rules, generates an ontology which can also be viewed as a syntactic equivalent of an RDF graph. This RDF graph can be accessed in three ways, as described in Sect. 4.3: ETL, SPARQL or Linked Data (lines 1, 3 and 2 in Fig. 4.3, respectively). In ETL, the RDF graph can be stored in a file or in a persistent triple store and be accessed directly from there, while in SPARQL and Linked Data access modes, the RDF graph is generated on the fly from the contents residing in the database. In these modes, a SPARQL query or an HTTP request respectively are transformed to an appropriate SQL query. Finally, the mapping engine retrieves the results of this SQL query (not depicted in Fig. 4.3) and formulates a SPARQL solution response or an appropriate HTTP response, depending on the occasion.

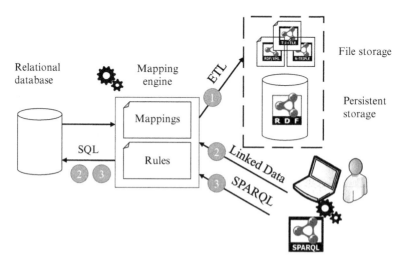

Fig. 4.3 Generation of ontology from a relational database (Spanos et al. 2012)

Before we delve into the different categories of ontology creation approaches, we describe as briefly as possible an elementary transformation method of a relational database to an RDF graph proposed by Tim Berners-Lee (Berners-Lee 1998) that, albeit simple, has served as a reference point for several ontology creation methods. This method has been further extended (Beckett and Grant 2003) so as to produce an RDFS ontology and is also informally known as "table-to-class, column-to-predicate" method, but for brevity we are going to refer to it as the *basic approach* throughout this Section. According to the basic approach:

(a) Every relation R maps to an RDFS class $C(R)$.
(b) Every tuple of a relation R maps to an RDF node of type $C(R)$.
(c) Every attribute *att* of a relation maps to an RDF property $P(att)$.
(d) For every tuple $R[t]$, the value of an attribute *att* maps to a value of the property $P(att)$ for the node corresponding to the tuple $R[t]$.

As subjects and predicates of RDF statements must be identified with a URI, it is important to come up with a URI generation scheme that guarantees uniqueness for distinct database schema elements. As expected, there is not a universal URI generation scheme, but most of them follow a hierarchical pattern, much like the one presented in Table 4.2, where for convenience multi-column primary keys are not considered. An important property that the URI generation scheme must possess and is indispensable for SPARQL and Linked Data access is reversibility, i.e., every generated URI should contain enough information to recognize the database element or value it identifies.

In Table 4.2, the name of the database is denoted *db*, *rel* refers to the name of a relation, *attr* is the name of an attribute and *pk* is the name of the primary key of a relation, while *pkval* is the value of the primary key for a given tuple of a relation.

The basic approach is generic enough to be applied to any relational database instance and to a large degree automatic, since it does not require any input from the human user, except from the base URI. However, it accounts for a very crude export of relational databases to RDFS ontologies, which does not permit more complex mappings nor does it support more expressive ontology languages. Moreover, the resulting ontology looks a lot like a copy of the database relational schema as all relations are translated to RDFS classes, even the ones which are mere artifacts of

Table 4.2 Typical URI generation scheme in a database-to-ontology mapping approach

Database element	URI Template	Example
Database	{*base_URI*}/{*db*}	http://www.example.org/company_db
Relation	{*base_URI*}/{*db*}/ {*rel*}	http://www.example.org/company_db/emp
Attribute	{*base_URI*}/{*db*}/ {*rel*}#{*attr*}	http://www.example.org/company_db/emp#name
Tuple	{*base_URI*}/{*db*}/ {*rel*}/{*pk=pkval*}	http://www.example.org/company_db/emp/id=5

the logical database design (e.g., relations representing a many-to-many relationship between two entities). Other notable shortcomings of the basic approach include its URI generation mechanism, where new URIs are minted even when there are existing URIs fitting for the purpose and the excessive use of literal values, which seriously degrade the quality of the RDF graph and complicate the process of linking it with other RDF graphs (Byrne 2008).

Nevertheless, as already mentioned, several methods refine and build on the basic approach, by allowing for more complex mappings and discovering domain semantics hidden in the database structure.

4.4.1   Creating and Populating a Domain Ontology

As mentioned before, approaches that create a "database schema ontology" are hardly useful for publishing Linked Data, since that ontology needs to be aligned with a domain ontology. These domain-specific ontologies do not contain any concepts or relationships that are related to the relational or entity-relationship models but instead, concepts and relationships pertaining to the domain of interest that is described by a relational database instance. Naturally, the expressiveness of the generated ontology largely depends on the amount of domain knowledge incorporated in the procedure. The two main sources of domain knowledge are the human user and the relational database instance itself. Therefore, there is a distinction between approaches that mainly rely on reverse engineering of the relational schema for the generation of a domain-specific ontology (but can also accept input from a human user) and approaches that do not perform extensive schema analysis but instead are mainly based on the basic approach and optionally, on input from a human expert. The latter seem to be more popular and have spawned more tools than the former, since they allow the human user to have full control over the definition of the mapping.

The automation level for such tools varies and depends on the level of involvement of the human user in the process. Most of these tools support all of the three possible workflows: from the fully automated standard basic approach with or without inspection and fine-tuning of the final result by the human user to manually defined mappings. Regarding data accessibility, all three modes have been adopted by several of the reviewed approaches with a strong inclination towards SPARQL based data access.

The language used for the representation of the mapping is particularly relevant for tools of this category. Unlike the case of tools outputting a database schema ontology, where the correspondences with database elements are obvious, concepts and relationships of a domain-specific ontology may correspond to arbitrarily complex database expressions. To express these correspondences, a rich mapping representation language that contains the necessary features to cover real world use cases is needed. Typical mapping representation languages are able to specify sets of data from the database as well as transformation on these sets that, eventually, define the form of the resulting RDF graph. Until the recommendation of R2RML by W3C, every tool used to devise its own native mapping language, each with unique syntax and features. This resulted in the locking of mappings that were created with a specific tool and could not be freely exchanged between different parties and reused across different mapping engines. It was this multilingualism that has propelled the development of R2RML as a common language for expressing database-to-ontology mappings.

Since the majority of the tools of this category rely on the basic approach for the generation of an ontology, the output ontology language is usually simple RDFS. The goal of these tools is rather to generate a lightweight ontology that possibly reuses terms from other vocabularies for increased semantic interoperability, than to create a highly expressive ontology structure. Vocabulary reuse is possible in the case of manually defined mappings, but the obvious drawback is the fact that, in order to select the appropriate ontology terms that better describe the semantics of the database contents, the user should also be familiar with popular Semantic Web vocabularies. The main motivation that is driving the tools of this Section is the mass generation of RDF data from existing large quantities of data residing in relational databases, which will in turn allow for easier integration with other heterogeneous data.

D2RQ (Bizer and Seaborne 2004) is one of the most prominent tools in the field of relational database-to-ontology mapping. It supports both automatic and user-assisted operation modes. In the automatic mode, an RDFS ontology is generated according to the rules of the basic approach and additional rules, common among reverse engineering methodologies, for the translation of foreign keys to object properties and the identification of M:N (many-to-many) relationships. In the manual mode, the contents of the relational database are exported to an RDF graph, according to mappings specified in a custom mapping language also expressed in RDF. A semi-automatic mode is also possible, in case the user builds on the automatically generated mapping in order to modify it at will. D2RQ's mapping language offers great flexibility, since it allows for mapping virtually any subset of a relation in an ontology class and several useful features, such as specification of the URI generation mechanism for ontology individuals or definition of translation schemes for database values. D2RQ supports both ETL and query-based access to the data. Therefore, it can serve as a programming interface and as a gateway that offers ontology-based data access to the contents of a database through either Semantic Web browsers or SPARQL endpoints. The engine that uses D2RQ mappings to translate requests from external applications to SQL queries on the relational database is called D2R Server (Bizer and Cyganiak 2006). Vocabulary reuse is also supported in D2RQ through manual editing of the mapping representation files.

OpenLink Virtuoso Universal Server is an integration platform that comes in commercial and open-source flavours and offers an RDF view over a relational database with its **RDF Views** feature (Blakeley 2007), which offers similar functionality to D2RQ. That is, it supports both automatic and manual operation modes. In the former, an RDFS ontology is created following the basic approach, while in the latter, a mapping expressed in the proprietary Virtuoso Meta-Schema language is manually defined. This mapping can cover complex mapping cases as well, since

Virtuoso's mapping language has the same expressiveness to D2RQ's, allowing to assign any subset of a relation to an RDFS class and to define the pattern of the generated URIs. ETL, SPARQL based and Linked Data access modes are supported. One downside of both D2RQ and Virtuoso RDF Views is the fact that a user should learn these mapping languages in order to perform the desired transformation of data into RDF, unless he chooses to apply the automatic basic approach.

Triplify (Auer et al. 2009) is another notable RDF extraction tool from relational schemas. SQL queries are used to select subsets of the database contents and map them to URIs of ontology classes and properties. The mapping is stored in a configuration file, the modification of which can be performed manually and later enriched so as to reuse terms from existing popular vocabularies. The generated RDF triples can be either materialized or published as Linked Data, thus allowing for dynamic access. Triplify also accommodates for external crawling engines, which ideally would want to know the update time of published RDF statements. To achieve this, Triplify also publishes update logs in RDF that contain all RDF resources that have been updated during a specific time period, which can be configured depending on the frequency of updates. The popularity of the Triplify tool is due to the fact that it offers predefined configurations for the (rather stable) database schemas used by popular Web applications. An additional advantage is the use of SQL query for the mapping representation, which does not require users to learn a new custom mapping language.

Ultrawrap (Sequeda and Miranker 2013) is a commercial tool that acts as a wrapper of a relational database as a SPARQL endpoint. It supports the creation of a new domain ontology by reverse engineering the database schema in order to obtain the initial conceptual model of the database. As the majority of approaches that extract a domain ontology from a relational schema, Ultrawrap applies a set of heuristic rules (Tirmizi et al. 2008) for the recognition of entities, attributes and relationships of a conceptual model, which forms the basis of the newly developed ontology. Ultrawrap offers SPARQL-based access to the relational data by rewriting a SPARQL query that contains terms from the newly developed domain ontology to an equivalent SQL one. Essentially, an RDF view is defined on top of the relational schema and incoming SPARQL queries are transformed to equivalent SQL ones operating on that RDF view. Therefore, the rewriting process in Ultrawrap is significantly simpler than typical SPARQL-to-SQL algorithms (Chebotko et al. 2009), (Cyganiak 2005), (Elliott et al. 2009), (Lu et al. 2007) mainly relying on the performance of the database optimization engine. Like D2RQ and Virtuoso RDF Views, Ultrawrap can also function with manually defined mappings that reuse terms from existing vocabularies.

Oracle DBMS also contains an RDF Views feature,[1] similar to that of OpenLink Virtuoso. The Oracle RDF Views functionality allows querying relational data, as if they were RDF graphs, therefore leading to the avoidance of physical storage for

---

[1] RDF Views in Oracle DBMS: docs.oracle.com/database/121/RDFRM/sem_relational_views.htm#RDFRM555

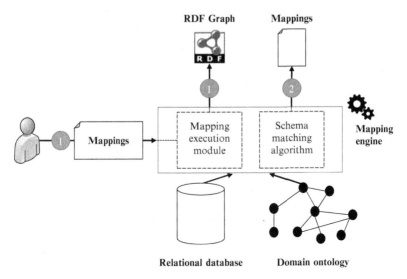

Fig. 4.4 Matching a relational database with one or more existing ontologies (Spanos et al. 2012)

RDF graphs. This feature allows automatic translation of a relational schema according to the Direct Mapping specification (mentioned in Sect. 2.6) as well as the definition of RDF graphs according to user-defined R2RML mappings. Furthermore, an interesting feature of Oracle DBMS is the ability to combine virtual and materialized RDF data in a query, by means of a special proprietary query operator that can incorporate the results of a SPARQL query in a SQL one.

4.4.2 Mapping a database to an existing ontology

In the previous Section, we discussed tools that do not require a priori knowledge of existing ontologies. On the contrary, there are also approaches that take as granted the existence of one or more ontologies and either try to establish mappings between a relational database and these ontologies or exploit manually defined mappings for generating an RDF graph that uses terms from these ontologies. These two alternatives are depicted in Fig. 4.4 (lines 2 and 1 respectively).

The underlying assumption adopted by all approaches is that the selected ontologies model the same domain as the one modeled by the relational database schema. Therefore, on the one hand, there are approaches that apply schema matching algorithms or augment reverse engineering techniques with linguistic similarity measures, in order to discover mappings between a relational database and an appropriately selected ontology. These mappings can then be exploited in various contexts and applications, such as heterogeneous database integration or even database schema matching (An et al. 2007), (Dragut and Lawrence 2004). On the other hand, there are tools that share quite a lot with certain tools from Sect. 4.4.1, receiving user-defined mappings between the database schema and one or more existing ontologies and generating an RDF graph that essentially populates ontologies with instance data from the database. In this Section, we are dealing with tools of the latter kind, since they are the ones that can be used directly for Linked Data publication.

Ontop (Rodriguez-Muro et al. 2013) is a promising tool that exploits manually defined mappings between a relational database schema and existing ontologies, in order to convert a relational database instance to a SPARQL endpoint. Ontop is a multi-purpose ontology-based data access framework that, besides simple SPARQL query answering, allows for application of the RDFS and OWL 2 QL entailment regimes. Practically, this means that facts that are inferred from the combination of the ontology and the relational data (the TBox and ABox respectively, in Description Logics parlance) need not be materialized but can be taken into account during query time. Ontop uses a sophisticated SPARQL-to-SQL rewriting algorithm that uses Datalog as an intermediate representation language and introduces several optimizations that produce as simple as possible SQL queries. The Ontop framework also includes a plugin for popular ontology editor Protégé (presented in Sect. 3.3.1.1), that enhances the latter with ontology-based data access functionality.

R_2O (Barrasa et al. 2004) is a declarative mapping language that allows a domain expert to define complex mappings between a relational database instance and one or more existing ontologies. R_2O is an XML-based language that permits the mapping of an arbitrary set of data from the database—as it contains features that can approximate an SQL query—to an ontology class. R_2O's expressiveness is fairly rich, supporting features such as conditional mappings or specification of URI generation schemes. The ODEMapster engine (Barrasa-Rodriguez and Gómez-Pérez 2006) exploits mappings expressed in R_2O in order to create ontology individuals from the contents of the database allowing for either materialized or query-driven access. Its successor, Morph (Priyatna et al. 2014) supports R2RML mappings, offering SPARQL-based data access to the contents of a database.

R2RML Parser (Konstantinou et al. 2014a), (Konstantinou et al. 2013a) is an open-source tool that can export RDF graphs from the contents of a relational database, based on an input R2RML mapping file. Its main components are a parsing engine, which analyzes an R2RML mapping file and materializes the specified RDF graph and a faceted browser of the generated RDF graph. Provided that R2RML Parser does not support dynamic access to relational database instances via SPARQL, it tackles the data synchronization issue by incorporating a feature of incremental dump (Konstantinou et al. 2014b). According to the latter, instead of generating from scratch the entire RDF graph, R2RML Parser updates accordingly the already dumped graph, based on updates made on the relational database. Section 4.5.2 showcases the tool operation in an example from the scholarly/cultural heritage domain.

Table 4.3 summarizes the tools presented in the current Section, including values for all relevant descriptive parameters introduced in Sect. 4.3.

Table 4.3 Tools for publishing linked data from relational databases

Tool	Level of automation	Data accessibility	Mapping language	Graphical user interface
D2RQ	Automatic/Manual	ETL/SPARQL/Linked Data	custom	No
Virtuoso RDF Views	Automatic/Manual	ETL/SPARQL/Linked Data	Virtuoso Meta-Schema Language/R2RML	Yes
Triplify	Manual	ETL/Linked Data	SQL	Yes
Ultrawrap	Automatic/Manual	SPARQL	R2RML	No
Ontop	Manual	ETL/SPARQL/Linked Data	custom/R2RML	Yes
R_2O/Morph	Manual	ETL/SPARQL	R_2O/R2RML	No
R2RML Parser	Manual	ETL	R2RML	No

4.5 Complete Example: Linked Data from the Scholarly/Cultural Heritage Domain

As far as the scholarly/cultural heritage domain is concerned, rich experience has led to the creation of software systems that demonstrate flawless performance over maintaining bibliographic archives. These systems operate at a high level of accuracy and fulfill their purpose. Therefore, one could argue that there is no significant reason for evolution, while it is also uncertain to which direction this evolution should take place. In order, however, to keep up with an ever-changing environment, regarding data and knowledge description, adoption of newer technologies is a necessity. This is firstly in order for the solutions to remain competitive but also, because of the benefits new technologies entail. Regarding digital repositories in this domain, the LOD paradigm offers efficient solutions to the following problems:

- *Integration*, typically materialized using OAI-PMH harvesting interfaces. The problem is that these interfaces, while often implemented and used in the digital libraries domain, do not ease integration with data models from other domains of discourse.
- *Expressiveness* in describing the information. OAI-PMH for instance, as far as information description is concerned, allows for a tree structure that extends to a depth-level of two. On the contrary, RDF allows for a graph-based description that can practically cover much more sufficiently the needs for describing information in the digital libraries domain.
- *Query answering* can be realized using conventional systems, to the extent where standard keyword-based queries are evaluated efficiently and respective results can span several repositories, offering a basic solution. However, querying graphs using graph patterns allows for much more complex queries, enabling formation of more descriptive queries.

Therefore, we can safely deduce that the benefits (query expressiveness, inherent semantics, and integration with third party sources) outweigh the disadvantages (resources investment in creating and maintaining the data). Because of these, and many other reasons as well, we can observe that more and more institutions open their data. For instance, the National Libraries in Spain (Biblioteca Nacional De Espana 2012), (Deutsche National Bibliothek 2012), (British Library 2012) are exposing free data services according to the LOD paradigm.

It is argued that Linked Data adoption is the future for digital repositories. A Linked Data compliant digital repository enables content re-use, allows participation of individual collections to the evolving global Linked Data cloud and gives the user the chance to discover new data sources at runtime by following data-level links, and thus deliver more complete answers as new data sources appear on the Web (Stevenson 2011).

When exposing a new collection as Linked Data, it is a good practice to reuse existing vocabularies/ontologies for its description as this makes it easier for the outside world to integrate the new data with already existing datasets and services (Knoth et al. 2011). Specifically for scholarly work, several vocabularies have been proposed. Table 4.4 gathers a number of ontologies developed only for the scholarly information domain, in order to indicate the wealth of existing and ongoing work in the domain.

A number of important aggregators with international coverage and diverse scope have made an appearance the scene in the last few years, such as the European digital heritage gateway Europeana,[2] DRIVER[3] and OpenAIRE[4] repositories of peer-reviewed scientific publications. Compatibility with aggregators is nowadays a task of high importance for repositories. It has become a common requirement for repositories to make their metadata records available for aggregators to retrieve, meeting specific criteria, and of course, adopting specific vocabularies. In this context, LOD adoption is the prevailing approach in the domain, bringing an order to the chaos of disparate solutions.

Overall, it turns out that the semantic technologies are an excellent solution for the domain. Exposing metadata as LOD can have a series of benefits, because of the following reasons:

- It allows avoidance of vendor lock-ins. In case there is a need for the mapping result to be migrated or moved to another tool, one can easily switch, i.e., parse the R2RML document using another tool such as Virtuoso.
- It allows complex queries to be evaluated on the results, utilizing the full capacities of SPARQL.
- Content can be harvested and integrated by third-party remote software clients in order to create valuable meta-search repositories such as Europeana or OpenAIRE,

---

[2] Europeana: www.europeana.eu

[3] DRIVER: www.driver-repository.eu

[4] OpenAIRE: www.openaire.eu

Table 4.4 Ontologies related to scholarly information

Title	URL	Name space	Namespace URL
The Bibliographic Ontology	bibliontology.com	bibo	http://purl.org/ontology/bibo/
Creative Commons Rights Ontology	creativecommons.org	cc	http://creativecommons.org/ns#
CiTo, the Citation Typing Ontology	purl.org/spar/cito	cito	http://purl.org/spar/cito/
Legacy Dublin Core element set	dublincore.org/documents/dces/	dc	http://purl.org/dc/elements/1.1/
DCMI Metadata Terms	dublincore.org/documents/dcmi-terms/	dcterms	http://purl.org/dc/terms/
FaBiO: FRBR-aligned bibliographic ontology	purl.org/spar/fabio	fabio	http://purl.org/spar/fabio/
FRBRcore	purl.org/vocab/frbr/core	frbr	http://purl.org/vocab/frbr/core#
FRBRextended	purl.org/vocab/frbr/extended#	frbre	http://purl.org/vocab/frbr/extended#
IFLA's FRBRer Model	iflastandards.info/ns/fr/frbr/frbrer/	frbrer	http://iflastandards.info/ns/fr/frbr/frbrer/
International Standard Bibliographic Description (ISBD)	iflastandards.info/ns/isbd/elements/	isbd	http://iflastandards.info/ns/isbd/elements/
Lexvo.org Ontology	lexvo.org/ontology	lvont	http://lexvo.org/ontology#
MARC Code List for Relators	id.loc.gov/vocabulary/relators	mrel	http://id.loc.gov/vocabulary/relators/
Open Provenance Model Vocabulary	purl.org/net/opmv/ns	opmv	http://purl.org/net/opmv/ns#
PRISM: Publishing Requirements for Industry Standard Metadata	prismstandard.org	prism	http://prismstandard.org/namespaces/basic/2.0/
Provenance Vocabulary Core Ontology	purl.org/net/provenance/ns	prv	http://purl.org/net/provenance/ns#
RDA Relationships for Works, Expressions, Manifestations, Items	rdvocab.info/RDARelationshipsWEMI	rdarel	http://rdvocab.info/RDARelationshipsWEMI
Schema.org	schema.org	schema	http://schema.org/

through which researchers can browse, search and retrieve content related to their work, such as scientific publications.

• Bringing existing content into the semantic web opens new capabilities about its migration, especially in the case where metadata is of larger volume than the actual data.

4.5.1   Synchronous vs. Asynchronous Exports as LOD in Digital Repositories

It must be noted that, as far as it concerns the choice between synchronous or asynchronous SPARQL-to-SQL translation in the digital repositories domain, the second option seems more viable. The cost of not having real-time results may not be as critical, considering that RDF updates could take place in a manner similar to maintaining search indexes. The trade-off in data freshness is largely remedied by the improvement in the query answering mechanism. Therefore, data freshness can be sacrificed in order to obtain much faster results (Konstantinou et al. 2013b). In the digital repositories domain, the choice of exposing data periodically comes at a low cost, because the information in this domain does not change as frequently as it does in other domains such as e.g., in sensor data, as discussed in Chap. 5.

Thus, although the asynchronous export approach from a relational database to RDF could be followed in many domains, it finds ideal application in the scholarly literature domain and, in general, in cases when data is not updated to a significant amount daily (Konstantinou et al. 2014a). In these cases, when freshness is not crucial and/or selection queries over the contents are more frequent than the updates, what our approach succeeds in, is a lower downtime of the system that hosts the generated RDF graph and serves query-answering.

4.5.2   From DSpace to Europeana: A Use Case

In this Section, we take a look at an example use case from the cultural heritage domain to allow scholarly use, using the Dublin Core model and the Europeana data model, the latter being a relatively flat metadata schema.[5] Let us consider, for instance, a repository with cultural heritage content. In the following example, it is shown how an item record has to be transformed in order to be exposed as RDF, in the EDM model.

The basic information flow in the proposed approach has the relational database as a starting point and an RDF graph as the result. The basic components are: the

---

[5] Europeana Data Model: pro.europeana.eu/edm-documentation

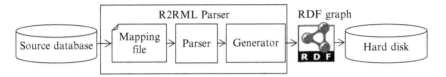

Fig. 4.5 Overall high-level architectural overview and information flow in the approach (Konstantinou et al. 2014a)

	Metadata field	Metadata value
Table 4.5 Example bibliographic record	dc.creator	G.C. Zalidis
		A. Mantzavelas
		E. Fitoka
	dc.title	Wetland habitat mapping
	dc.publisher	Greek Biotope-Wetland Centre
	dc.date	1995
	dc.coverage.spatial	Thermi
	dc.type	Article
	dc.rights	http://creativecommons.org/licenses/by/4.0/

source relational database, the R2RML Parser tool,[6] and the resulting RDF graph. Figure 4.5 illustrates how this flow in information processing takes place. First, database contents are parsed into result sets. Then, according to the mapping file, defined in R2RML (see Sect. 2.6), the Parser component generates a set of instructions (i.e., a Java object) for the Generator component. Subsequently, the Generator component, based on this set of instructions, instantiates in-memory the resulting RDF graph. Next, the generated RDF graph is persisted. Note that using Jena, this could be an RDF file on the hard disk or TDB, Jena's custom implementation of threaded B+ Trees, introduced in Sect. 3.4.3.

Table 4.5 above gives a subset of an example bibliographic record, as it is stored in a digital repository platform. The example is based on an item whose persistent identifier (URI) is http://hdl.handle.net/11340/615.

The goal is to expose the above as an RDF description, taking into account the Europeana Data Model and the Dublin Core metadata set. Next, we provide a snippet of the output description, in abbreviated RDF/XML syntax. This can be achieved with use of the R2RML Parser (introduced in Sect. 4.4.2), and R2RML (see Sect. 2.6) mapping declarations.

---

[6]R2RML Parser: www.w3.org/2001/sw/wiki/R2RML_Parser

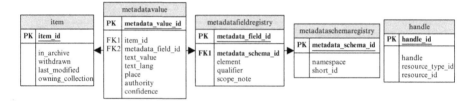

Fig. 4.6 Metadata value storage in DSpace

```
<edm:ProvidedCHO
rdf:about="http://www.example.org/handle/11340/615">
  <dc:creator
rdf:resource="http://www.example.org/persons#G.C. Zalidis"/>
  <dc:creator rdf:resource="http://www.example.org/persons#A.
Mantzavelas"/>
  <dc:creator rdf:resource="http://www.example.org/persons#E.
Fitoka"/>
  <dc:title>
    Wetland habitat mapping
  </dc:title>
  <dc:publisher
rdf:resource="http://www.example.org/publishers#Greek
Biotope-Wetland Centre"/>
  <dc:date>
    1995
  </dc:date>
  <dcterms:spatial
rdf:resource="http://www.example.org/spatial_terms#Thermi"/>
  <dc:type
rdf:resource="http://www.example.org/types#Article"/>
  <dc:rights>
    http://creativecommons.org/licenses/by/4.0/
  </dc:rights>
</edm:ProvidedCHO>
```

For the relational database schema of DSpace,[7] which is a leading institutional repository solution, this can take place using R2RML mappings for each metadata field. Internally, DSpace provides a basic infrastructure, on which metadata can be stored, following arbitrary schemas and vocabularies. Let us see next how metadata values are stored in the relational database supporting a DSpace instance: Figure 4.6 illustrates the respective part of the DSpace database model: table `metadatavalue` holds all metadata values for all items. Each metadata value belongs to a metadata field, registered in the `metadatafieldregistry` table that stores the metadata fields for each schema that, in its turn, exists in the `metadataschemaregistry` table. We note that PK stands for Primary Key, while FK1 and FK2 refer to Foreign Keys.

---

[7] DSpace: www.dspace.org

Based on this model, we can construct the R2RML map, SQL queries and respective RDF classes to populate. In the following, we present a triples map definition that will create URIs based on metadata values from DSpace. Specifically, in the example below, metadata values coming from the metadata field dc.coverage.spatial will be converted to RDF triples having:

- **Subject** according to the rr:subjectMap template (`'http://www.example.org/handle/{"handle"}'`)
- **Predicate** according to the rr:predicate value (dcterms:spatial)
- **Object** according to the rr:objectMap template (`'http://www.example.org/spatial_terms#{"text_value"}'`)

The example above is not complete, as it covers only one metadata field, aiming at showing one of the steps of the transformation process. Similar correspondences can be accordingly defined for the rest of the metadata fields of the repository schema. Finally, note that the triples generated by this example contain links to another resources that may be organized in a different graph, as the objects in the produced triples are subject URIs in the spatial terms graph, instead of being just literals.

```
map:dc-coverage-spatial
   rr:logicalTable <#dc-coverage-spatial-view>;
   rr:subjectMap [
      rr:template 'http://www.example.org/handle/{"handle"}';
   ];
   rr:predicateObjectMap [
      rr:predicate dcterms:spatial;
      rr:objectMap [
         rr:template
         'http://www.example.org/spatial_terms#{"text_value"}';
         rr:termType rr:IRI
      ];
   ].

<#dc-coverage-spatial-view>
   rr:sqlQuery """
   SELECT h.handle AS handle, mv.text_value AS text_value
   FROM handle AS h, item AS i, metadatavalue AS mv,
metadataschemaregistry AS msr, metadatafieldregistry AS mfr
WHERE
   i.in_archive=TRUE AND h.resource_id=i.item_id AND
   h.resource_type_id=2 AND
   msr.metadata_schema_id=mfr.metadata_schema_id AND
   mfr.metadata_field_id=mv.metadata_field_id AND
   mv.text_value is not null AND i.item_id=mv.item_id AND
   msr.namespace='http://dublincore.org/documents/dcmi-terms/'
   AND mfr.element='coverage' AND mfr.qualifier='spatial'
   """.
```

It has to be underlined that the technical dimension of exporting data as RDF is half of the problem of publishing repository data: the second half concerns its bibliographic dimension; and is a whole project by itself. As noted several times in this book, widespread ontologies have to be used where applicable in order to offer meaningful, semantically enriched descriptions of the repository data. Moreover, linking the data to third party datasets, using other datasets' identifiers is also an aspect that is discussed in Sects. 3.3.1 and 3.4.2.3.

4.6 Future Outlook

The problem of extracting RDF graphs from relational databases has been extensively investigated and has led to the development of several tools that are approaching a stable and mature level. Nevertheless, there still exist some issues and challenges that have not been yet efficiently encountered and will be the subject of research in the years to follow. The most significant ones are:

Ontology-based data update. A lot of approaches mentioned offer SPARQL-based access to the contents of the database. However, this access is unidirectional. Since the emergence of SPARQL Update that allows update operations on an RDF graph, the idea of issuing SPARQL Update requests that will be transformed to appropriate SQL statements and executed on the underlying relational database has become more and more popular. Some early work has already appeared in the clj-r2rml (Garrote and Moreno García 2011) and OntoAccess (Hert et al. 2011) prototypes, as well as the D2RQ/Update (Eisenberg and Kanza 2012) and D2RQ++ (Ramanujam et al. 2010) extensions of the D2RQ tool. However, these approaches suffer from various drawbacks each: for example, D2RQ++ uses a secondary triple store during the update process, while D2RQ/Update does not follow strictly the semantics of SPARQL Update. Moreover, most approaches consider only basic (relation-to-class and attribute-to-property) mappings. The issue of updating relational data through SPARQL Update is similar to the classic database view update problem, therefore porting already proposed solutions would contribute significantly in dealing with this issue.

Mapping update. Database schemas and ontologies constantly evolve to suit the changing application and user needs. Therefore, established mappings between the two should also evolve, instead of being redefined or rediscovered from scratch. This issue is closely related to the previous one, since modifications in either participating model do not simply incur adaptations to the mapping but also cause some necessary changes to the other model as well. So far, only few solutions have been proposed for the case of the unidirectional propagation of database schema changes to a generated ontology (Curé and Squelbut 2005) and the consequent adaptation of the mapping (An et al. 2008). The inverse direction, i.e., modification of the database as a result of changes in the ontology has not been investigated thoroughly yet. On a practical note, both database trigger functions and mechanisms like the Link

Maintenance Protocol (WOD-LMP) from the Silk framework (Volz et al. 2009) could prove useful for solutions to this issue.

Generation of Linked Data. Reusing terms from popular Semantic Web vocabularies in a database-to-ontology mapping is not sufficient for the generation of RDF graphs that can be easily interlinked with other datasets in the LOD cloud. For the generation of true Linked Data, the real-world entities that database values represent should be identified and links between them should be established, in contrast with the majority of current methods, which translate database values to RDF literals. Lately, a few interesting tools that handle the transformation of spreadsheets to Linked RDF data by analyzing the content of spreadsheet tables have been presented, with the most notable examples being the RDF extension for Google Refine (see Sect. 3.4.2.1) and T2LD (Mulwad et al. 2010). Techniques as the ones applied in these tools can certainly be adapted to the relational database paradigm.

References

An Y, Borgida A, Mylopoulos J (2006) Discovering the semantics of relational tables through mappings. J Data Semantics 7:1–32

An Y, Borgida A, Miller RJ et al (2007) A semantic approach to discovering schema mapping expressions. In: Chirkova R, Oria V (eds) Proceedings of 2007 IEEE 23rd international conference on data engineering (ICDE 2007), Istanbul, Turkey, April 2007. IEEE, pp 206–215

An Y, Hu X, Song IY (2008) Round-trip engineering for maintaining conceptual-relational mappings. In: Bellahsène Z, Léonard M (eds) Advanced information systems engineering: 20th international conference (CAiSE 2008), Montpellier, France, June 2008. Lecture notes in computer science, vol. 5074. Springer, Heidelberg, pp 296–311

Auer S, Dietzold S, Lehmann J et al (2009) Triplify—Light-weight linked data publication from relational databases. In: WWW'09, Proceedings of the 18th international conference on World Wide Web, New York, NY, USA, pp 621–630

Barrasa J, Corcho O, Gómez-Pérez A (2004) R2O, an extensible and semantically based database-to-ontology mapping language. Second international workshop on semantic web and databases (SWDB 2004), 29–30 August 2004, Toronto, Canada

Barrasa-Rodriguez J, Gómez-Pérez A (2006) Upgrading relational legacy data to the semantic web. In: WWW'06, Proceedings of the 15th international conference on World Wide Web, Edinburgh, Scotland, April 2006. ACM, New York, pp 1069–1070

Beckett D, Grant J (2003) SWAD-Europe Deliverable 10.2: mapping semantic web data with RDBMSes. http://www.w3.org/2001/sw/Europe/reports/scalable_rdbms_mapping_report/. Accessed 26 Dec 2014

Ben Necib C, Freytag JC (2005) Semantic query transformation using ontologies. In: Desai BC, Vossen G (eds) Proceedings of 9th international database engineering & application symposium (IDEAS 2005), Montreal, Canada, July 2005. IEEE, pp 187–199

Berners-Lee T (1998) Relational databases on the semantic Web. http://www.w3.org/DesignIssues/RDB-RDF.html. Accessed 26 Dec 2014.

Biblioteca Nacional De Espana (2012) Linked data at Spanish National Library. http://www.bne.es/en/Catalogos/DatosEnlazados/index.html. Accessed 26 Dec 2014

Bizer C, Seaborne A (2004) D2RQ—Treating non-RDF databases as virtual RDF graphs. Poster presented at the 3rd international semantic web conference (ISWC 2004), Hiroshima, Japan

Bizer C, Cyganiak R (2006) D2R Server—Publishing relational databases on the semantic web. Poster presented at the 5th international semantic web conference (ISWC 2006), Athens, Georgia, USA

Blakeley C (2007) Virtuoso RDF views getting started guide. http://www.openlinksw.co.uk/virtuoso/Whitepapers/pdf/Virtuoso_SQL_to_RDF_Mapping.pdf. Accessed 26 Dec 2014

British Library (2012) Linked data at the British Library. http://www.bl.uk/bibliographic/datafree.html. Accessed 26 Dec 2014

Buccella A, Penabad MR, Rodriguez FR et al (2004) From relational databases to OWL ontologies. In: Proceedings of 6th Russian conference on digital libraries (RCDL 2004), Pushchino, Russia

Byrne K (2008) Having triplets—holding cultural data as RDF. In: Larson M, Fernie K, Oomen J et al (eds) Proceedings of the ECDL 2008 workshop on information access to cultural heritage. Aarhus, Denmark

Chebotko A, Lu S, Fotouhi F (2009) Semantics preserving SPARQL-to-SQL translation. Data & Knowledge Eng 68(10):973–1000

Curé O, Squelbut R (2005) A database trigger strategy to maintain knowledge bases developed via data migration. In: Bento C, Cardoso A, Dias G (eds) Progress in artificial intelligence: 12th Portuguese conference on artificial intelligence (EPIA 2005), Covilha, Portugal, December 2005. Lecture notes in computer science, vol 3808. Springer, Heidelberg, pp 206–217

Cyganiak R (2005) A relational algebra for SPARQL. Hewlett Packard. http://www.hpl.hp.com/techreports/2005/HPL-2005-170.pdf. Accessed 26 Dec 2014

Das S, Srinivasan J (2009) Database technologies for RDF. In: Tessaris S, Franconi E, Eiter T et al (eds.) Reasoning web. Semantic technologies for information systems: 5th international summer school 2009. Lecture notes in computer science, vol 5689. Springer, Heidelberg, pp 205–221

Deutsche National Bibliothek (2012) The linked data service of the German National Library. http://www.dnb.de/EN/Service/DigitaleDienste/LinkedData/linkeddata_node.html. Accessed 26 Dec 2014

Dragut E, Lawrence R (2004) Composing mappings between schemas using a reference ontology. In: Meersman R, Tari Z (eds) On the move to meaningful internet systems 2004: CoopIS, DOA, and ODBASE. Lecture notes in computer science, vol 3290. Springer, Heidelberg, pp 783–800

Eisenberg V, Kanza Y (2012) D2RQ/Update: updating relational data via virtual RDF. In: Proceedings of the 21st international conference companion on World Wide Web (WWW'12 Companion), Lyon, France, April 2012. ACM, New York, pp 497–498

Elliott B, Cheng E, Thomas-Ogbuji C et al (2009) A complete translation from SPARQL into efficient SQL. In: Desai BC (ed) Proceedings of the thirteenth international database engineering & applications symposium (IDEAS'09), Calabria, Italy, September 2009. CM, New York, USA, pp 31–42

Elmasri R, Navathe SB (2010) Fundamentals of database systems. The Benjamin/Cummings, San Francisco

Garrote A, Moreno García MN (2011) RESTful writable APIs for the web of linked data using relational storage solutions. In: Bizer C, Heath T, Berners-Lee T et al (eds) Proceedings of the 4th linked data on the web wWorkshop (LDOW 2011), Hyderabad, India. CEUR Workshop Proceedings, vol 813

Geller J, Chun SA, An YJ (2008) Toward the semantic deep web. Computer 41(9):95–97

Ghawi R, Cullot N (2007) Database-to-ontology mapping generation for semantic interoperability. Paper presented at the 3rd international workshop on database interoperability (InterDB 2007), held in conjunction with VLDB 2007, Vienna

Gómez-Pérez A, Corcho-Garcia O, Fernandez-Lopez M (2003) Ontological engineering. Springer-Verlag, New York

Heath T, Bizer C (2011) Linked data: evolving the web into a global data space. Morgan & Claypool, San Rafael

Hellmann S, Unbehauen J, Zaveri A et al (2011) Report on knowledge extraction from structured sources. LOD2 Project. http://static.lod2.eu/Deliverables/deliverable-3.1.1.pdf. Accessed 26 Dec 2014

Hendler J (2008) Web 3.0: chicken farms on the semantic web. IEEE Computer 41(1):106–108

Hert M, Ghezzi G, Würsch M et al (2011) How to "Make a Bridge to the New Town" using ontoaccess. In: Aroyo L, Welty C, Alani H et al (eds) The semantic web—ISWC 2011, Proceedings of the 10th international semantic web conference, part II, Bonn, Germany, October 2011. Lecture notes in computer science, vol 7032. Springer, Heidelberg, pp 112–127

Knoth P, Robotka V, Zdrahal Z (2011) Connecting repositories in the open access domain using text mining and semantic data. In: Gradmann S, Borri F, Meghini C et al (eds) Research and advanced technology for digital libraries, Proceedings of the international conference on theory and practice of digital libraries 2011 (TPDL 2011), Berlin, Germany, September 2011. Lecture notes in computer science, vol 6966. Springer, Heidelberg, pp 483–487

Konstantinou N, Spanos DE, Mitrou N (2008) Ontology and database mapping: a survey of current implementations and future directions. J Web Eng 7(1):1–24

Konstantinou N, Spanos DE, Stavrou P et al (2010) Technically approaching the semantic web bottleneck. Int J Web Eng Technol 6(1):83–111

Konstantinou N, Houssos N, Manta A (2013) Exposing bibliographic information as linked open data using standards-based mappings: methodology and results. 3rd international conference on integrated information (IC-ININFO '13), Elsevier, Prague, Czech Republic

Konstantinou N, Spanos DE, Mitrou N (2013) Transient and persistent RDF views over relational databases in the context of digital repositories. In: Garoufallou E, Greenberg J (eds) Metadata and semantics research. Proceedings of the 7th MTSR conference, Thessaloniki, Greece, November 2013. Communications in Computer and Information Science, vol 390. Springer, Heidelberg, pp 342–354

Konstantinou N, Spanos DE, Houssos N et al (2014a) Exposing scholarly information as linked open data: RDFizing DSpace contents. Electronic Library 32(6):834–851

Konstantinou N, Kouis D, Mitrou N (2014b) Incremental export of relational database contents into RDF graphs. In: Akerkar R, Bassiliades N, Davies J et al (eds) Proceedings of the 4th international conference on web intelligence, mining and semantics (WIMS '14), June 2014, Thessaloniki, Greece. ACM, New York

Lu J, Cao F, Ma L et al (2007) An effective SPARQL support over relational databases. In: Christophides V, Collard M, Gutierrez C (eds) Semantic web, ontologies and databases: VLDB workshop (SWDB-ODBIS 2007), Vienna, Austria, September 2007. Lecture notes in computer science, vol 5005. Springer, Heidelberg, pp 57–76

Mulwad V, Finin T, Syed Z et al (2010) Using linked data to interpret tables. In: Hartig O, Harth A, Sequeda J (eds) Proceedings of the first international workshop on consuming linked data (COLD 2010), Shanghai, China, November 2010. CEUR workshop proceedings. vol 667

Poggi A, Lembo D, Calvanese D et al (2008) Linking data to ontologies. J Data Semantics 10:133–173

Priyatna F, Corcho O, Sequeda J (2014) Formalisation and experiences of R2RML-based SPARQL to SQL query translation using morph. In: Proceedings of the 23rd international conference on World Wide Web (WWW'14), Seoul, Republic of Korea. ACM, New York, USA, pp 479–490

Ramanujam S, Khadilkar V, Khan L et al (2010) Update-enabled triplification of relational data into virtual RDF stores. Int J Semantic Comput 4(4):423–451

Rodriguez-Muro M, Kontchakov R, Zakharyaschev M (2013) Ontology-based data access: ontop of databases. In: Alani H, Kagal L, Fokoue A et al (eds) The semantic web—ISWC 2013, Proceedings of the 12th international semantic web conference, October 2013, Sydney, Australia. Lecture notes in computer science, vol 8218. Springer, Heidelberg, pp 558–573

Sahoo S, Halb W, Hellmann S et al (2009) A survey of current approaches for mapping of relational databases to RDF. World wide web consortium. http://www.w3.org/2005/Incubator/rdb2rdf/RDB2RDF_SurveyReport.pdf. Accessed 26 Dec 2014

Sequeda JF, Tirmizi SH, Corcho O et al (2009) Direct mapping SQL databases to the semantic web: a survey. University of Texas, Austin, Department of Computer Sciences. ftp://ftp.cs.utexas.edu/pub/techreports/tr09-04.pdf. Accessed 26 Dec 2014

Sequeda JF, Miranker DP (2013) Ultrawrap: SPARQL execution on relational data. J Web Semantics 22:19–39

Sheth AP, Larson JA (1990) Federated database systems for managing distributed, heterogeneous, and autonomous databases. ACM Computing Surveys 22(3):183–236

Spanos DE, Stavrou P, Mitrou N (2012) Bringing relational databases into the semantic web: a survey. Semantic Web J 3(2):169–209

Stevenson A (2011) Linked data—the future for open repositories? Presentation at the open repositories (OR 2011) conference, Austin, TX

Tirmizi SH, Sequeda JF, Miranker DP (2008) Translating SQL applications to the semantic web. In: Bhowmick SS, Küng J, Wagner R (eds) Database and expert systems applications: 19th international conference (DEXA 2008), Turin, Italy, September 2008. Lecture notes in computer science, vol 5181. Springer, Heidelberg, pp 450–464

Volz J, Bizer C, Gaedke M et al (2009) Discovering and maintaining links on the web of data. In: Bernstein A, Karger DR, Heath T et al (eds.) The semantic web—ISWC 2009: Proceedings of the 8th international semantic web conference, Washington DC, October 2009. Lecture notes in computer science, vol 5823. Springer, Heidelberg, pp 650–665

Volz R, Handschuh S, Staab S et al (2004) Unveiling the hidden bride: deep annotation for mapping and migrating legacy data to the semantic web. J Web Semantics 1(2):187–206

Wache H, Vögele T, Visser U et al (2001) Ontology-based integration of information—a survey of existing approaches. In: Gómez-Pérez A, Gruninger M, Stuckenschmidt H et al (eds) Proceedings of the IJCAI-01 workshop on ontologies and information sharing, Seattle, USA, August 2001. CEUR Workshop Proceedings,, vol 47, pp 108–117

Zhao S, Chang E (2007) From database to semantic web ontology: an overview. In: Meersman R, Tari Z, Herrero P (eds) On the move to meaningful internet systems: OTM 2007 Workshops. Lecture notes in computer science, vol 4806. Springer, Heidelberg, pp 1205–1214

Chapter 5
Generating Linked Data in Real-time from Sensor Data Streams

5.1 Introduction: Problem Framework

During the past few years, a rapid evolution in ubiquitous technologies has occurred, leaving back conventional desktop computing and advancing to the era where pervasive computing is part of everyday experience. Computer-based devices are seen supporting various activities, based either on user input or on information sensed by the environment. This evolution, in combination with a parallel decrease of the price of sensors has recently allowed sensor applications to make their appearance in a variety of domains. As a consequence, nowadays, more than ever, we are witnessing the materialization of visions and concepts such as the Internet-of-Things (IoT) (Sundmaeker et al. 2010) and M2M communications (Lawton 2004).

Therefore, as streamed data proliferates at increasing rates, the need for real-time, large-scale stream processing application deployments increases as well. In the same way that data stream management systems (DSMS) have emerged from the database community, there is now a similar concern in managing dynamic knowledge among the Semantic Web community.

Numerous challenges are associated with such efforts. The large scale and the geographic dispersion of such environments as well as the volume of data, the multiple distributed heterogeneous components that need to be assembled (e.g., spanning sensor, sensor processing and signal processing (including audiovisual) components) and the diversity of their vendors are some of the issues that need to be addressed. Additionally, the need for automation, since manual observation of multiple sensor (e.g., camera) feeds is not possible, and the inclusion of high-level intelligent reasoning for event detection and inference are features without which the added value of the system is limited.

These needs originally led to the development of several DSMS (Arasu et al. 2004), (Chandrasekaran et al. 2003), (Abadi et al. 2005), so as to fill the gap left by traditional Database Management Systems, which are not geared towards dealing with continuous, real-time sequences of data. DSMS introduced a novel approach

© Springer International Publishing Switzerland 2015
N. Konstantinou, D.-E. Spanos, *Materializing the Web of Linked Data*,
DOI 10.1007/978-3-319-16074-0_5

that is not based on persistent storage of all available data and user-invoked queries, but instead follows an approach of on-the-fly stream manipulation and permanent monitoring queries.

In this Chapter, we will see how we can employ techniques and standards described in previous Chapters, in order to semantically annotate streamed data. The Chapter is structured as follows: Section 5.2 introduces the basic concepts of context-awareness, Internet of Things and describes how the Linked Data domain fits in the picture. Section 5.3 deals with the notion of fusion, which is essential in dealing with large volumes of information. Section 5.4 focuses on the data layer: on the data that is stored locally in the nodes that comprise the system and the ways they are collected, processed and stored and communicated in the system. In Sect. 5.5, rule-based stream reasoning is discussed while Sect. 5.6 presents a practical approach on how a Linked Data repository can be populated with data originating from a multi-sensor fusion system that collects data in real-time.

5.2   Context-Awareness, Internet of Things, and Linked Data

Context-aware systems are the systems that have knowledge of the environment in which they are acting and they behave differently, according to the existing conditions. The term "context-awareness" is tightly connected to the Internet of Things (IoT) vision, which aims at connecting (large numbers of) sensors deployed around the world. In the IoT, the focus shifts to automated configuration of filtering, fusion and reasoning mechanisms that can be applied to the collected sensor data streams, originating from selected sensors.

According to (Dey 2001), information related to the environment is any information that can be used to characterize the state of an entity. An entity can be a person, a place, an object, a system user, or even the system itself. A system is considered to be "context-aware" when it is in position to extract and use information from the environment in order to adjust its functionality according to the incoming information.

Because of rapid evolutions in the domain, various concepts have been developed and studied by researchers: ubiquitous, pervasive and context-aware systems in this Section are terms that refer more or less to the same property: The ability of a system to "read" its environment and take advantage of this information in its behavior.

The challenge in these systems lies in the fact that the systems that capture, store, process, and represent information are usually heterogeneous, while the information is generated in large volumes and at a high velocity. A common representation format can serve the purpose of making the systems interoperable, thus allowing it to be shared, communicated and processed by several systems.

The Linked Data ecosystem offers an efficient approach towards filling in this gap. Evolution in description and query languages, storing and sharing information led to the creation of a common, reliable and flexible framework for information management. Without a certain level of information homogeneity, a system comprising heterogeneous components (e.g., consisting of a set of cameras and microphones and targeting at monitoring a place) can be quickly obsolete as it will be monolithic and very likely hard to use. Additionally, using formalisms with inadequately defined semantics make information integration a very difficult task (Lassila and Khushraj 2005).

5.3 Fusion

Information fusion is the study of techniques that combine and merge information and data residing at disparate sources, in order to achieve improved accuracies and more specific inferences than could be achieved by the use of a single data source alone (Llinas and Waltz 1990; Sheth and Larson 1990). Fusion implies a leverage of information's meaning while in the same time a partial loss of initial data may occur. This loss does not necessarily degrade the value of the information conveyed. Contrarily, the resulting fused values will convey additional value compared to the initial ones.

Fusion can take place at many levels, especially in multi-sensor systems, regarding how close the fused information is to the actual data stream. Broadly, the levels at which fusion can be considered as taking place include the following:

- *Signal level*. When signals are received simultaneously by a number of sensors, the fusion of these signals may lead to a signal with a better signal-to-noise ratio.
- *Feature level*. In this level, a component comes into play, a so-called "perceptual" component, with the task to extract the desired low-level features from each modality and represent them, typically in a multidimensional vector. Such examples are presented in (Li et al. 2003) and (Wu et al. 2004) where normalization and transformation matrices are used in order to derive a single "multimodal" feature vector that will be subsequently used during analysis. However, the abstraction at the feature level may lead to loss of possibly supporting information since the perception of the real world cannot usually be complete with the use of a single medium.
- *Decision level*. Fusion at this level involves combining information from multiple algorithms in order to yield a final fused decision, and may be defined by specific decision rules. Fusion at this level is often termed *late fusion* as opposed to *early fusion*, taking place at the Signal and Feature levels.

Another important fusion property concerns the algorithm perception of the world. Consider a fusion system consisted of a sensor network that collects information. Using an online (or distributed) algorithm, each node can take decisions based only on its perception of the world. In this case, each algorithm execution is based on the knowledge of only a local node or a cluster of nodes. Contrarily, in offline (or centralized) algorithms there is a need of a central entity maintaining system-wide information.

Fusion nodes are nodes whose purpose is to process data coming from other nodes or sensors. They can operate in one of the two modes:

- Act as a server, waiting to be contacted. The node can then be considered as operating in *push* mode in the sense that information is *pushed* from the sensors to the fusion node.
- Harvest information that is collected in the distributed nodes or sensors. This mode of operation is referred to as *pull* mode.

It is noted that *information integration* is a concept fundamentally different than fusion. Compared to integration, we could state that information fusion takes place in the processing steps while integration refers to the final step: the end user's gateway to (integrated) access to the information.

5.3.1 JDL Fusion Levels

Sensor fusion and context awareness are fields in which the military has shown special interest. Thus, in order to improve communications among military researchers and system developers, the Joint Directors of Laboratories (JDL) Data Fusion Working Group began an effort to define the terminology related to data fusion. The result of that effort was the creation of a process model for data fusion and a data fusion lexicon (Hall and Llinas 2001). The JDL process model is a paper model of data fusion and is intended to be very general and useful across multiple application areas. The JDL data fusion process model is a conceptual model which identifies the processes, functions, categories of techniques, and specific techniques applicable to data fusion (see Fig. 5.1). According to the model data fusion process is conceptualized by sensor inputs, human-computer interaction, database management, source preprocessing, and four key subprocesses:

- *Level 1 (Object Refinement)*: is aimed at combining sensor data together to obtain a reliable estimation of an entity position, velocity, attributes, and identity;
- *Level 2 (Situation Refinement)*: dynamically attempts to develop a description of current relationships among entities and events in the context of their environment;

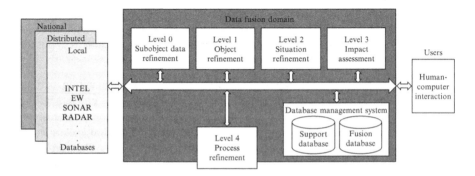

Fig. 5.1 Fusion levels according to the JDL notion (Hall and Llinas 2009)

- *Level 3* (*Threat Refinement*): projects the current situation into the future to draw inferences about enemy threats, friendly and enemy vulnerabilities, and opportunities for operations;
- *Level 4* (*Process Refinement*): is a meta-process, which monitors the overall data fusion process to assess and improve the real-time system performance.

Revisions of the model suggest several modifications such as the addition of a higher Level 5 (*Cognitive* or *User Refinement*), which introduces the human user in the fusion loop where the aim is to generate fusion information according to the needs of the system user. This information can include for instance visualization of the fusion products and generation of feedback or controls to enhance/improve these products (Blasch and Plano 2003).

The JDL model designed to describe data fusion offers a generally accepted categorization of the levels at which fusion can take place. However, even this distinction among levels is not concrete, and thus the JDL model appears in the bibliography as having four or five levels according to the context.

5.4 The Data Layer

In Sect. 1.2.1, a brief introduction was made distinguishing the terms data and information. Also in the context of data stream systems, the terms are subjective and vary according to the context. For instance, when analyzing video streams in order to extract meaningful information, our data could be the video frames while the information could be a generated annotation. Our data, in the context of this Section are in fact metadata of multimedia (incl. audiovisual) streams, in a semi-structured form, originating from a set of heterogeneous distributed sources. Using a middleware, this data flows into the system and is semantically annotated, forming a Knowledge Base, able to answer queries at a higher, semantic level.

As noted in Sect. 1.2.4, *semi-structured* data sources are sources that—as opposed to *structured*—carry their structure within the data. In this sense, relational databases contain structured data, as the database schema is stored separately from the data, while XML files are semi-structured, as the structure is inseparable from the data. With this in mind, and keeping in mind that RDF itself carries in the same document the (graph) structure and the (graph) data, the problem of converting simple semi-structured documents (e.g. in XML or JSON) into RDF poses no significant technological challenges as simple transformations suffice.

Data is produced in the form of streams. The difference between data streams and archives is that in the case of streams, there is no starting or ending point. Therefore, processing in both cases is identical, with the difference that in the case of streamed data, a strategy must be devised in order to define how new facts will be pushed into the system and old facts will be pushed out of it.

We describe next, how this data is modeled, how annotations based on the annotation model are generated, stored, and communicated throughout the system.

5.4.1 Modeling Context

Context refers to *"any information that can be used to characterize the situation of an entity. An entity is a person, place, or object that is considered relevant to the interaction between a user and an application, including the user and applications themselves"* (Dey 2001). Modeling context, a system's perception of the world, is hardly a trivial task. The challenge in modeling context is in the complexity of capturing, representing and processing the concepts that comprise it. The description of the world must be expressive enough in order to enable specific behaviors but not as complex as to render the collected information unmanageable.

A common representation format and vocabulary should be followed especially in the case when contextual information needs to be transferred among applications, in order to ensure syntactic and semantic interoperability, while enabling integration with third party data sources. Also, uniform context representation and processing at the infrastructure level enables better reuse of derived context by multiple data producers and consumers. Ontology-based descriptions help toward this direction since they offer unambiguous definitions of the concepts and their relationships. In addition, such approaches allow further exploitation of the created Knowledge Base through the possibility of submitting and evaluating higher level intelligent semantic queries.

Therefore, semantic representation of sensory data is significant because ontologies can be used in order to specify the important concepts in a domain of interest and their relationships, thus formalizing knowledge about the specific domain. In association with semantic annotation, ontologies and rules play an important role in the Semantic Sensor Web vision for interoperability, analysis, and reasoning over heterogeneous multimodal sensor data.

With these in mind, several approaches have been proposed in the literature, which apply Semantic Web technologies on top of a sensor network in order to support the services provided in the existing and the newly deployed sensor networks. These approaches aim at covering descriptions of the most widespread concepts in the domain and apply technologies that allow the sensor data to be understood and processed in a meaningful way by a variety of applications with different purposes. Ontologies are used for the definition and the description of the semantics of the sensor data.

Employment of Semantic Web technologies in sensor networks research focuses mainly on sensor modeling in order to enable higher level processing for event/situation analysis. Initiatives such as (Eid et al. 2007) use SUMO as their core ontology under which they define a sensor ontology for annotating the sensors and it can be then queried for sensor discovery. In this approach, the authors claim that among the benefits of using this ontology is the maximization of the precision of searching sensor data by utilizing semantic information.

Other frameworks, such as the one proposed in (Neuhaus and Compton 2009), define ontologies for sensor measurement and sensor description. The framework emerges from the W3C Semantic Sensor Network Incubator Group[1] which aims at providing ontologies and semantic annotations that define capabilities of sensors and sensor networks.

The authors in (Toninelli et al. 2006) follow an approach based on describing the knowledge base model using *resource, actor, environment* as its base concepts. As the authors describe, information stored in a context model can be logically organized in:

- parts that describe the state of the resources (the resource part)
- the entities operating in the resources (the actor part), and
- the surrounding environment conditions (the environment part).

The ontology (the main information gathering point) is defined according to the domain where the sensor network is placed and it defines the entities that are taking place in a situation, the events and the relations between them.

In order to take advantage of research that has been conducted in the area of semantic enabled situation awareness, the Situation Theory Ontology (STO) (Kokar et al. 2009) has been developed. In STO, the authors used an ontology to create a unified expression of the Situation Theory (Barwise 1981; Endsley 2000). The STO is expressed in OWL, which enables situations to be described using a formal language, thus allowing inference through a reasoning engine or by the use of appropriate rules. STO models the events/objects and their relationships in a way that can be extended using either OWL axioms and properties or rules for supporting complex cause-effect relations that cannot be expressed in OWL alone. The STO can be extended with classes and relations that correspond to the actual application scenario. In order to be able to use it in a real sensor fusion environment two additional ontologies are integrated. These are the Time ontology[2] which holds the timestamp of any concept instance that is stored during runtime and the WGS84 Geo Positioning ontology[3] which holds the latitude and longitude values of entities.

Other attempts include the SSN ontology (Lefort et al. 2011), suggested by the W3C Semantic Sensor Network Incubator Group, which attempts to become a standardized approach for describing sensor data in the "Linked Sensor Data" context, bridging the Internet of Things and the Internet of Services. Going one step further, the Semantic Sensor Web (SSW) is proposed as a framework for providing enhanced meaning for sensor observations so as to enable situation awareness (Sheth et al. 2008). This is accomplished through the addition of semantic annotations to existing standard sensor languages of OGC's www.opengeospatial.org Sensor Web Enablement. These annotations provide more meaningful descriptions and enhanced access to sensor data in comparison with the Sensor Web Enablement alone, and they act as a linking mechanism to bridge the gap between the primarily syntactic XML-based metadata standards of the Sensor Web Enablement and the RDF/OWL-based vocabularies driving the Semantic Web.

---

[1] Semantic Sensor Network Incubator Group: www.w3.org/2005/Incubator/ssn/

[2] Time Ontology in OWL: www.w3.org/TR/owl-time/

[3] Basic Geo (WGS84 lat/long) Vocabulary: www.w3.org/2003/01/geo/

Overall, in the case of a sensor fusion system where new types of sensors need to be integrated and there is a relatively high variety in its structure, a Local-As-View setting (see Sect. 1.2.2) is considered optimal. In other words, there should be a global schema to which every sensor node should conform. This approach is preferred to translating user queries from the global schema to each one of the local ones in order to retrieve, aggregate and return results to the user. All data that are produced by the sensors and are communicated in the system are eventually stored in its Knowledge Base, where reasoning can be performed in order to infer knowledge which corresponds to events, and situation assessment that cannot be performed at lower fusion levels.

5.4.2   Annotation of Sensor Data

Sensor data, in general, can be regarded to as belonging in one of two major categories: audiovisual and non-audiovisual data. In the former case, the sensor data that is produced is essentially an audio/video stream while in the latter it can be any kind of streamed sensor observations (temperature, RFID tags, etc.). In both cases incoming data is primarily subject to filtering. This means that not all information is stored and processed but only the subset that is meaningful and needed by the processing modules. For instance, values that are out of range will be discarded and only data that meets certain conditions can be acquired for processing.

In order to sense and capture data from the environment, sensors and appropriate software is needed. The sensors in essence convert the analog signals received from the environment into sensory data that in their raw form are usable to the sensor driver only. In order to elevate the meaning of the sensed information and allow its transfer and reuse by other applications, a common representation format should be followed, enabling a common information basis. Moreover, in order to allow syntactic—and additionally semantic—interoperability, the annotation of the various potentially heterogeneous data sources is a task as crucial as challenging.

Before proceeding with more detailed description of how sensor data annotation can take place, we should keep in mind that sensor data annotation does not constitute a single panacea for its retrieval and direct utilization by applications. There are several reasons for that. First, there is no single approach as far as homogenizing bulks of data. Numerous standards offer the means to annotate various areas of sensor data such as the MPEG-7 for audiovisual, Federal Geographic Data Committee (FGDC) standards for geographical information etc. Second, it remains as a question how accurate the annotations are, how convenient it is to update and maintain them in accurate state and, mostly important, what the true added value to the content itself is. In other words it is not clear how can the user benefit from the existence of such annotations.

Moreover, problems such as false annotations, missing or incorrect annotations that can be produced by tracker errors have to be dealt with. However, in large collections of sensor data, imperfect annotation is preferred to no annotation at all. Approaches can be broadly regarded as consisting of two steps:

- First, sensor data is annotated according to the nature of its tracker.
- Second, this data is homogenized under a common vocabulary that contains both the values and the semantics captured by the overall system deployment.

5.4.3   Real-time vs. Near-real-time, Synchronous vs. Asynchronous

Especially in multi-sensor systems, real-time information processing is an important issue, as the processing system/middleware has to be scalable in order to be able to offer answers/solutions in real-time. However, real-time processing is not always possible or desirable.

A *real-time* system is one that must satisfy explicit bounded response time constraints to avoid failure and present consistency regarding the results and the process time needed to produce them. The fundamental difference between a real-time and a non real-time system is the emphasis in predicting the response time and the effort in reducing it. In (Dougherty and Laplante 1995), a system's *response time* is defined as "the time between the presentation of a set of inputs and the appearance of all the associated outputs". Therefore, in a sensor fusion system that, for instance, produces certain alerts, the notion of real-time dictates that these alerts (outputs) are connected to the appearance of certain inputs (events).

Systems that schedule their operations at fixed time intervals cannot be considered real-time. *Near real-time* is a characterization that can be applied if these intervals are frequent enough. For a system that employs semantic web technologies, this means that unless the entire Knowledge Base is reloaded and reasoning procedures are performed on it, we cannot consider that all the associated outputs are produced and the system cannot be classified as real-time. The frequency of updates depends each time on the application. For instance, in a surveillance scenario, more frequent updates would be needed than in an environmental monitoring scenario.

More specifically, when new facts are appearing in the system, one can choose whether to just store them and process them at a later time, asynchronously, or to calculate the changes reflected at the entire Knowledge Base at once. The former approach is not real-time, although it can produce near real-time results if the calculations are performed frequently enough. The latter is the real-time approach and would perform optimally at low data incoming rates or small computational overhead.

Considering each sensor as a system with inputs and outputs, it can produce either real-time signals (e.g., audiovisual streams) or near-real time/asynchronous messages, e.g., RFID readings, taking place at fixed time intervals.

5.4.4   Data Synchronization and Timestamping

Synchronization is an important decision that needs to be taken when designing our time management strategy. The approach followed in timestamping the generated messages and events may be not as trivial as it seems. A system for fusing information might have rules of the form: if *event₁* occurred before *event₂* then...

In such cases, deviations from a common clock may lead to irregularities. If we consider a fusion node in a fusion graph, the events that are generated can have two kinds of timestamps:

- The time the event was recognized: the local time in the node that measured the property. In this case, the event is timestamped close to its actual creation, which is ideal since it reflects the real world conditions. The problem that arises in this case is how to synchronize the sensor network to a common clock.
- The time the event arrived in the fusion node (a node that fuses information incoming from other nodes). This revokes the need of a common clock, since the fusion node will timestamp events upon their arrival. However, this approach should be followed when there are no great delays in communicating messages between nodes.

Additionally, time measurements can be made either at each node (distributed) or at a central node, maintaining a common clock (centralized).

5.4.5   Windowing

In streaming, regardless to whether the streamed data is generated at high or low rates, a mechanism needs to be developed to assure continuous data processing. For the newly generated information, it needs to be assured that in order to process it properly, the system will not have to take into account all existing information. In other words, it is important that the system maintains a working memory window since, by definition, streams are unbounded and cannot fit into memory in order to be processed as a whole. Regarding windowing, there are three important decisions that have to be taken:

- *The measurement unit.* The window is defined in certain units that can be for instance tuples in a relational database or triples in a RDF triple store. Units can be physical, as the aforementioned or logical, defined in terms of time (e.g., milliseconds) (Patroumpas and Sellis 2006). Time-based windowing, however, tends to produce more fluctuation in the response times (Konstantinou et al. 2010).
- *The size.* The window size increases proportionally the system latency because of the larger data volume. Therefore, a large window size in a fusion node, although it will reflect a more complete perception of the world, it may reduce its performance. Therefore, the window size has to be balanced according to the accepted latency. On one hand, having all the data available would be ideal but

on the other hand, resources in terms of processing capabilities and storage are limited and latency should always be kept at an acceptable level. This could be remedied with the adoption of hierarchical or incremental processing approaches (Barbieri et al. 2010), which enable taking into account the whole window but focus only at the latest points.

- *The window behavior.* Memory windows, depending on how they are updated, they can be categorized as follows:

  - *Sliding windows* have a fixed length and both their endpoints are moving at a predefined interval, replacing old stream elements with new ones.
  - *Tumbling windows* can be considered as a special case of sliding windows, where the window is moving at a fixed interval of length equal to the window size. Thus, tumbling windows are pairwise-disjoint and do not share any stream elements. This behavior, although convenient for a large number of cases, it is not always optimal since window moving operations could take place at inconvenient times (e.g., during the occurrence of an event of interest) leading to incorrect results (Konstantinou et al. 2010).
  - *Landmark windows* maintain a fixed beginning endpoint while the other one is moving, resulting in a variable window size (Arasu et al. 2006).
  - *Partitioned windows* dictate that windows are formed according to attributes of interest (Li et al. 2005).
  - *Predicate windows*, a model introduced in (Ghanem et al. 2006), limit the stream tuples to the focus of a continuous query, in a manner similar to the partitioned windows.

Regardless of the windowing behavior, older elements that are dropped out of the window are discarded, leading to an obvious loss of information. This imposes restrictions in the expressiveness of the rules that can be defined and which shape the system's behavior, i.e., the rules that process newly generated information. Because at all times only the latest information is available, we cannot declare rules that have to take into account older—discarded or archived—information. The rules applied to real-time are restricted to the current information window. Specifically, in order to state a rule that needs to take into account archived information, each rule execution would impose extra computational burden. Therefore, among the drawbacks of maintaining a window of information we have to include the reduction in the rule expressiveness in the window. Of course, this does not restrict the expressiveness of the rules that can process archived information.

5.4.6   The (Distributed) Data Storage Layer

In order to support the required capabilities, the information that is generated needs to be stored and processed before being communicated to the system. For this reason, each node and its associated information has to maintain its perception of the real world, physically stored in a local database.

Among the most important decisions that have to be taken while designing the databases that support the system is to balance between completeness of the view of the real-world facts in each database on one hand and system scalability on the other. Since a system designed for multi-sensor stream processing is purposed to function under a heavy load of information generated by numerous sensors, the database design that is mostly preferred is geared towards scalability (i.e., decentralized to the maximum extent possible) instead of centralizing the collected information.

Therefore, an approach in order to maintain scalability would be that each of the nodes keeps the amount of information required for its operation locally and communicates to the central node only higher level information which comprises e.g., detected events or entities. Consequently, because of the various sensor data processing components that comprise the architecture, providing to the sensor nodes and the upper layers of the infrastructure information that does not necessarily need to be interwoven, the schema of each node is depending on the components themselves. Using this approach, the local database schema in each sensor node has to relate only to the components it hosts, while the restriction imposed is in the information that is communicated throughout the system.

5.4.6.1 Leveraging Information Value: Relational to RDF in Sensor Data Streams

As explained in the previous Section, the data that are produced and fused close to the sensors, can be kept close to the sensors in order to avoid flooding the network with streaming messages. However, the produced messages will be fused, forwarded to the upper layers and ultimately inserted in an ontology. Therefore, they will eventually have to be translated at some point into semantic notations. This task can be performed by a mapping layer, which maps the relational schema to the semantic schema. There are two strategies for accomplishing this transformation, using either a push or a pull method (similarly to how the data are forwarded to fusion nodes, they can also be forwarded from the database to the semantic layer).

Push. The push method forwards the data to the ontology using semantic notation as soon as they are generated. This has to be implemented in the lower level with the semantic layer having a passive role in the process. The advantage of this method is that the transformations are executed fast and the ontology is always up-to-date. The disadvantages is that each lower level node will have to implement its own push method while there is the risk that the ontology will be populated with data even when no query is sent to the semantic layer.

Pull. The pull method transforms the relational data to semantic on request, i.e., during query time. This approach is similar to RDF Views (see Sect. 4.4.1) where mappings, simple or complex, between relational database tables and ontology concepts and properties are defined. During query time, the mapping process is triggered and data are transformed on the fly. The advantages of the pull method is that the actual mapping is defined at the higher semantic level rather than the lower levels and that data are transformed on request so that the ontology will accumulate instances that are needed for the actual query evaluation. The disadvantage of this method is that it could lead to longer response times during queries.

Additional information (even not originating from the sensors) can be integrated and fused at a higher level, in order to derive situations. This means that information from third party sources, as long as it is exposed via appropriate service interfaces, can be queried, collected and included in inference.

5.5 Rule-Based Stream Reasoning in Sensor Environments

Reasoning means inferring implicit information from explicitly asserted one. Inference, in general, is based on a set of rules that is provided and applied on the knowledge base that is created. Reasoning uses the ontology structure and the stored instances in order to draw conclusions about the ongoing situations. The knowledge base can further be extended by using rules to describe situations/events that are too complex to be defined using OWL notation only. It is important to note that sensor data captured by an ontology can be combined with a rule-based system that can reason over the ontology instances creating alarm-type of objects if certain conditions are met (Trifan et al. 2008).

The use of rules is essential in depicting the desired behaviour of context-aware systems. Since the model-theoretic background of the OWL language is based on Description Logics systems that are a subset of the First Order Logic, the designed model has fully defined semantics and, also, Horn clauses can be formed upon it. These clauses can be seen as rules that predefine the desired intelligence in the system's behavior.

In general, two distinct sets of rules can be applied: one containing rules that specify how sensor measurements represented in an XML-based format will be mapped to a selected ontology and another set of rules deriving new facts, actions or alerts based on existing facts and the underlying knowledge represented by the ontology. The first set can be referred to as mapping rules and the second one as semantic rules. The rules are formed according to the following event-condition-action pattern (on *event* if *condition* then *action*) (Papamarkos et al. 2003) where the event in a sensor network is a message arrival indicating a new available measurement. Mapping rules can fetch data from the XML-like message and store it into the ontology model in the form of class instances. They can be perceived as the necessary step bridging the gap between semi-structured data presented in an XML form and ontological models. Semantic rules, on the other hand, can perform modifications, when needed, solely on the ontology model. This set of rules depend on the semantics of the specific domain or deployment scenario and involves high-level concepts that are meaningful to humans e.g., "when a certain area under observation is too hot, open the ventilating system". Of course, the "too hot" conclusion will probably be inferred from a set of current observations coupled with the knowledge stored in the ontology.

The rules are fired based on data obtained from sensors in real-time and classified as instances of fundamental ontologies. The rule-based system is used as an advisor in the complex process of decision making, applying corrective measures when needed. The decision combines the real-time data with a-priori sensor data stored in the system. The platform controls also the whole environment in order to make the user aware of any glitches in the functionality of the whole system.

Rule-based problem solving has been an active topic in AI and expert systems over the last decades. The classic approach to rule-based computational systems comes from work on logical programming and deductive databases. Conventional rule engine implementations are based on the Rete algorithm (Doorenbos 1995).

The W3C recommendation that has been created for this purpose is based on the Rule Interchange Format (RIF) (see Sect. 2.7), a core rule language and a set of extensions (dialects) that allow the serialization and interchange of different rule formats. Other approaches include RuleML[4] and SWRL (Horrocks et al. 2004).

Another approach for the definition of a rule language is through the use of SPARQL CONSTRUCT clauses. An extension to this approach is SPIN (see Sect 2.7). Although this seems like a simple solution, CONSTRUCT queries that use the full features of the non-monotonic SPARQL language (especially the combination of left join and union, which is the main source of complexity (Perez et al. 2006)), result in an expressive and complex rule language (non-recursive Datalog with negation (Schenk 2007)), and its combination with ontologies needs to be further studied (Polleres 2007).

Additionally, while reasoners have excelled their performance with regard to static knowledge, the issue of reasoning with frequently changing facts and assertions has not been adequately addressed yet. *Stream reasoning* has been coined as a term that describes the above problem (Della Valle et al. 2008) of performing reasoning on a knowledge base comprising stable or occasionally changing terminological axioms and a stream of incoming assertions or facts. Advances and solutions to this problem will progressively lead the way for smarter and more complex applications, including for instance traffic management and fastest route planning, environmental monitoring, surveillance and object tracking scenarios, and disease outburst detection applications among others.

Next, in Sects. 5.5.1 and 5.5.2, we present how rule-based reasoning can be implemented in Jena and Virtuoso. Note that the capabilities offered by the Jena framework can be used in conjunction with Virtuoso, since a framework for their integration is offered natively by Virtuoso. This enables access to the Virtuoso RDF Quad store through Jena. Thus, Jena can cooperate with Virtuoso in order to allow development of applications.[5]

---

[4] RuleML: wiki.ruleml.org

[5] Virtuoso Jena Provider: www.openlinksw.com/OdbcRails/main/Main/VirtJenaProvider

5.5.1   Rule-Based Reasoning in Jena

The Jena Semantic Web Framework (introduced in Sect. 3.4.3) is the most popular Java framework for ontology manipulation. Apart from the rest of its rich capabilities, Jena can also be used as a reasoner, since it includes an internal inference engine. The framework includes a number of predefined reasoners (the Transitive reasoner, the RDFS rule reasoner, the OWL, OWL mini, OWL micro reasoners, the DAML micro reasoner) and, most importantly, includes a generic rule reasoner that can be customized to meet specific ad hoc application demands. Essentially, Jena reasoners are built-in rule files of the form of the following example:

```
[rdfs5a: (?a rdfs:subPropertyOf ?b), (?b rdfs:subProperty
Of ?c) -> (?a rdfs:subPropertyOf ?c)]
```

The example above is a forward chain rule (body → head) that states that when the body is true, then the head is also true. This type of rules is also used in order to support symmetric and transitive properties in OWL, as in the following examples, respectively:

```
[symmetricProperty1: (?P rdf:type owl:SymmetricProperty),
(?X ?P ?Y) -> (?Y ?P ?X)]
[transitiveProperyl: (?P rdf:type owl:TransitiveProperty),
(?A ?P ?B), (?B ?P ?C) -> (?A ?P ?C)]
```

Additionally, Jena supports a set of builtin primitives, functions that can be used in the place of rule predicates, such as *isLiteral(?x), notLiteral(?x), isFunctor(?x) notFunctor(?x), isBNode(?x), notBNode(?x)*. Notably, this set of primitives is extendable and allows implementation of custom builtin primitives that can be used in order to develop custom checks or actions, such as triggering an alert based on a triple set pattern.

5.5.2   Rule-Based Reasoning in Virtuoso

Virtuoso includes RDF inference capabilities, in order to infer triples that are not physically stored but are implied by the existing triples. The context upon which inference takes place can be built from one or more graphs containing RDFS triples. In order to perform reasoning, Virtuoso imports the supported RDFS or OWL constraints from these graphs and groups them together into rule bases. Rule bases, in Virtuoso, are persistent entities that can be referenced by SPARQL queries or endpoints. The queries that are evaluated against triple sets with a given rule set return results as if the triples inferred by the rule set were included in the graph (or graphs) accessed.

Reasoning in Virtuoso is essentially a set of rules applied on the initial RDF graph. The inferred triples—that occur after applying this rule set on the initial set of triples—are not physically stored in the Virtuoso storage engine but queries that are running with a given rule base (i.e., the rules that define how reasoning is executed). Reasoning procedures in Virtuoso are kept relatively simple, compared to the reasoning capabilities that OWL can offer. This approach allows Virtuoso to scale, instead of offering rich inference capabilities.

In order to load rule sets, Virtuoso offers the `rdfs_rule_set` function. Using this function, the user can specify a logical name for the rule set, plus a graph URI. A rule set can be used in order to combine multiple schema graphs into one single rule set. Additionally, a schema graph may also participate in multiple rule sets.

In order to use the rule set in the context of a query, the user first has to load the RDF (or OWL) schema into the Virtuoso triple store. After the document is loaded, the user can add the assertions into an inference context using the `rdfs_rule_set` function. Rules that apply to the reasoning are provided as a context to user queries. The extra triples that may be generated by the reasoning (the rule set) are then generated at runtime as needed. Thus, the triples that are entailed by e.g., a subclass or a subproperty statement in an inference context are not stored physically but are added to query results at runtime.

5.6   Complete Example: Linked Data from a Multi-Sensor Fusion System Based on GSN

In this Section we outline the fusion capabilities potentially offered by a system that fuses information originating from a distributed sensor network. As shown next, the solution comprises a multi-level fusion system, operating at all JDL (see Sect. 5.3.1) levels, it blends ontologies with low-level information databases, and seamlessly combines semantic web technologies with a sensor network middleware.

In the scope of a proof-of-concept implementation and also as a testbed for our experiments, the snapshot of the architecture above that was implemented employs one computer node that hosts a Low Level Fusion (LLF) Node and two processing components: a Smoke Detector and a Body Tracker (each one hosted on a computer with a camera that streams its feed using Real-time Transport Protocol (RTP)), a node that offers High Level Fusion (HLF) capabilities and, finally, a Central Node that has the overall system supervision.

5.6.1   The GSN Middleware

In order to perform LLF, the example relies on the Global Sensor Networks (GSN) middleware[6] that is an open-source, java-based implementation, designed in order to allow processing from a large number of sensors and as such, it covers the

---

[6] Global Sensor Networks (GSN) middleware: github.com/LSIR/gsn/wiki

functionality requirements of LLF in sensor data streams. In order to acquire information, GSN introduces the concept of *virtual sensor*. Any data provider, not only sensors, can provide data to a GSN instance, as long as a virtual sensor configuration file (in the form of an XML) defines the processing class, the sliding window size, the data source and the output fields. GSN Servers can communicate between them, thus forming a network where information is collected, communicated, fused and integrated in order to produce the desired results.

As far as it concerns data acquisition, each GSN node can support input from more than one data stream. In order to combine and fuse the information, an SQL-like procedural language is offered. This allows to the user to define LLF functions in the manner presented in the following section, while GSN takes care behind the scenes about crucial issues such as thread safety, synchronization, etc.

5.6.2   Low Level Fusion

Bottom-up, the system can be described as follows: First, the camera generates an RTP feed with its perception of the world. The feed is processed by both the signal processing components (the Body tracker and the Smoke detector). The first component generates a stream containing at all times the `NumberOfPersons` detected by the component. The second one, simple however, is not as straightforward: it splits the image into particles and reports the `NumberOfSmokeParticles` detected. Figure 5.2 illustrates the two components.

Both the components (the Body Tracker and the Smoke Detector) will be receive a `POLL` command at fixed time intervals and will return an XML containing their perception of the world: the number of persons tracked, and the number of smoke particles, respectively.

Note that sampling the video source takes place asynchronously. This happens because the camera will be streaming at 25 frames per second (fps), while for the

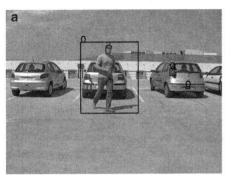

Fig. 5.2 Perception of the world, as tracked by two components. In (**a**) the BodyTracker component, based on the Camshift algorithm (Bradski 1998) and in (**b**) a Smoke Detector component, as analyzed in Avgerinakis et al. (2012)

a   **b**

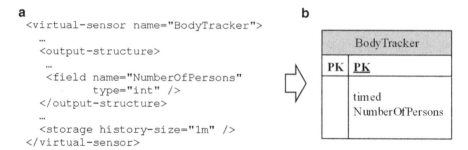

```
<virtual-sensor name="BodyTracker">
  ...
  <output-structure>
    ...
    <field name="NumberOfPersons"
           type="int" />
  </output-structure>
  ...
  <storage history-size="1m" />
</virtual-sensor>
```

Fig. 5.3 Using GSN, virtual sensor descriptions are translated into local relational database tables

needs of the LLF, 500 ms may suffice between two consecutive polls. In addition, tracking a person is a more demanding task than detecting it, since the former task implies comparing consecutive frames for differences between them, while for the latter task the processing of single frames is enough. Therefore, the processing behavior is characterized as asynchronous, since the camera fps rate is not aligned with the messages per second that the component produces.

Using GSN, we can define the measured properties by each virtual sensor. The virtual sensor declaration of a field of the name NumberOfPersons will be declared in the virtual sensor XML configuration file:

Similarly, the smoke detector can have an ExistenceOfSmoke variable, of Boolean type. These declarations will lead to the creation of a local database schema for each sensor, of the form of Fig. 5.3b, where the field name is populated with sensor measurements, while two fields are added by GSN: PK is an auto-incremented integer, being the primary key, and timed by applying a timestamp on the measurement. GSN takes care of the windowing, which can be configured in the virtual sensor configuration file, either as tuple-based, i.e., the number of tuples to keep in the local database, or time-based, i.e., keeping only the tuples that were generated in the defined period. The virtual sensor configuration in Fig. 5.3a defines a time window of one minute for the historical data stored locally. Next, a third, virtual sensor, the LLF Virtual sensor, will fuse the data, by applying queries of the following form, in its local database.

```
SELECT source1. NumberOfPersons AS v1,
source2. ExistenceOfSmoke AS v2
FROM source1, source2
WHERE v1>0 AND v2="true"
```

In the example above, there are two data providers (source streams) for one GSN Server, source1 and source2. This virtual sensor definition will produce events only in the case when the WHERE condition is satisfied, that is only when at least a person and smoke are detected, concurrently. The example above demonstrates the concept of fusion: results are provided by taking into account the inputs from both the sensors. This query will generate results only in the case when the fusion conditions are true. Of course, more processing components, sensors, sensor types and more complex fusion conditions can be added in the processing scheme.

So far, no semantic enrichment has been performed. However, using the above-mentioned approach, signal processing components as illustrated in Fig. 5.2, audio-visual or not, can be integrated into the system architecture and populate the system knowledge, as detailed next in Sect. 5.6.3. Low Level Fusion is necessary in order to allow only important information to be forwarded to the upper layers of the system and populate its knowledge base.

5.6.3 A Sensor Fusion Architecture

Fusion, in the scope of a multi-sensor system, focuses in the definition of a sensor fusion architecture which addresses matters that correspond to all JDL levels, from JDL level 0 to JDL level 4, i.e., from subobject data refinement to process refinement. The architecture in Fig. 5.4 builds on top of the generic architecture illustrated in Fig. 5.1.

With respect to this background, the functionality of this multi-sensor architecture can be categorized in the following discrete JDL levels:

- JDL Level 0: At this level, the system collects observation data, performs spatial and temporal alignment and virtualizes the inputs of the sensors. This is the level at which the Signal Processing components operate. The feature extraction components (such as cameras, microphones, etc.) operate on sensory data input mostly from Electro-Optical (EO) sensors.
- JDL Level 1: At this level the system analyzes the collected observation data provided by the sensors attached to the system, in order to extract features, objects, and low level events. For instance, the detected *objects* may include persons or vehicles, while low level *events* may include movements. This implementation can be based on e.g., a GSN system

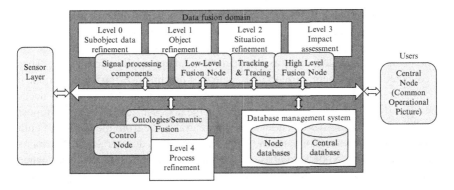

Fig. 5.4 Fusion Levels in a proposed multi-sensor system architecture (Doulaverakis et al. 2011)

- JDL Level 2: The system, at this level, aggregates and fuses the objects, events and the general context in order to refine the common operational environment. At this level, the system can perform object fusion and tracking.
- JDL Level 3: The events and objects that have been refined at JDL Level 2 are subsequently analyzed, using semantically enriched information, resulting at the inferred system state. High Level Fusion (HLF, e.g., using Virtuoso's rule engine) operates at this level.
- JDL Level 4: At this level, the system can refine its operation by providing feedback to the sensor layer, in order to improve the fusion process. This is realized by the Ontologies/Semantic fusion component that performs analysis on the system-wide high-level collected information in order to infer events and potential risks and threats. This component is hosted at the Central Node.
- A higher level, introduced in revised versions of the JDL model, Level 5 (Cognitive or User Refinement) in which the Control Node and Common Operational Picture components operate. These components are both hosted at the Central Node and they are responsible for issuing commands back to the sensors and providing a visualization of the system state with regard to the deployed sensors, detected objects, sensed events and inferred threats, respectively.

The above mentioned levels can be further understood by considering together Figs. 5.1 and 5.4, illustrating how we can move from theory to practice. As presented in Fig. 5.1, in order to leverage the data collected into meaningful and semantically enriched information, all levels of fusion are applied to the data, transforming it from signals to higher level information. In order to achieve the task of multi-sensor data fusion, we notice that first the data originates from the sensor layer and is first processed by the signal processing components (JDL Level 0/1). The LLF Nodes extract objects and situations (JDL Level 1/2). At the next fusion level, a Tracking and Tracing component can identify situations in the system, in a manner that is not feasible by a sensor alone (JDL Level 2). Next, the HLF Node, given the objects and situations detected, fuses the information, enabling reasoning that can infer and assess potential impacts (e.g., situations and respective alerts), at JDL Level 3. At Level 4, the semantic fusion in combination with the central capabilities of the system can monitor and curate the whole system, allowing for process refinement. Finally, at Level 5, the Central Node hosts the system front-end and allows for human-computer interaction.

The databases that support the system's operation are of two distinct types. The support database is materialized as a node database, kept locally at each node. Thus, the collected data are kept close to their source of origin, allowing for each node to be configured and maintained according to its environment and system's needs. Consequently, the choice of keeping the data in the nodes instead of a central repository allows the system to scale, alleviates the Central Node from the burden of communicating large volumes of data, and allows a system of a decentralized, distributed nature. The Central Node's database plays the role of the fusion database, where higher level fused information is kept.

5.6.4 High Level Fusion Example

Ontologies, as analyzed before, offer powerful means to describing concepts and their relations, and as such, they can be employed to provide enough specificity in describing the higher level concepts of JDL levels 2 and 3 but also to describe the relations between these concepts (Llinas et al. 2004). An ontology provides the standardized form in which situations are defined so that appropriate algorithms for context awareness can be formulated.

The ontology is the main information gathering point in the proposed architecture. It defines the entities that are taking place in a situation, the events and the relations between them. All data that are produced by the processing modules and by the LLF nodes are eventually stored here where reasoning is performed in order to infer new knowledge which corresponds to events and situation assessment that cannot be performed at the LLF level.

In order to take advantage of research that has been conducted in the area of semantic enabled situation awareness, the proposed solution uses as its core ontology the Situation Theory Ontology (STO) (see Sect. 5.4.1) which is based on Situation Theory.

Now, in order to demonstrate the inference capabilities of the proposed architecture, a scenario was set-up which makes full use of the data chain from low level sensor data to high level complex event detection and situation awareness/threat assessment. The example demonstrates how external services can be utilized in order to detect "critical" situations in an urban environment.

We selected video cameras which can be used to detect persons and also incidents, such as smoke, in an area of interest. Sensor locations are associated with a priori knowledge during reasoning. Detected smoke at places such as a petrol station mark a significant threat to public safety, based on a priori knowledge about the location and can be associated with a clear emergency action plan.

Smoke detection in critical location situation: Reasoning in Virtuoso requires that the underlying triple store was populated with inferred event data from the Smoke detector and Body tracker components (see Fig. 5.2). RDF views in Virtuoso have been developed and used to load the underlying data from the sensor inference database. Technically, this meant configuring GSN in order to use Virtuoso as its backend at the HLF node. Then, using the scheduler component in Virtuoso we trigger a procedural logic which updates the triple store via RDF views from the sensor inference database and subsequently invoke on the reasoning.

In the example of smoke detection at the petrol station the relevant smoke event together with location data of the video camera is forwarded to emergency personnel with the relevant information. The reasoning process associates the smoke detection event with a criticality factor according to the events and geospatial information modeled in STO. The process of modeling events in STO is described in detail in (Kokar et al. 2008). In this example, we used the `STO:FocalSituation` class to mark significant situations that prompt action by security personnel. The class

```
SELECT ?focalsituation ?timeTxt ?locTxt
WHERE {
   ?focalsituation STO:focalRelation ?event .
   ?event STO:hasAttribute ?time .
   ?event STO:hasAttribute ?location .
   ?time rdf:type STO:Time .
   ?time time:inXSDDateTime ?timeTxt .
   ?location rdf:type STO:Location .
   ?location rdfs:label ?locTxt .
}
```

Fig. 5.5 SPARQL Query example, retrieving "significant situations" as modeled using STO

interlinks the data associated with events which we can query using SPARQL and forward the results (e.g. to an emergency responsible) via a Web service call.

The query in Fig 5.5 is run against the system model and retrieves focal situations with their event related details such as time location and further textual descriptions. Finally, integration with emergency departments can be achieved by sending details on significant threats found in SPARQL result sets to a Web service endpoint exposed in an emergency department.

References

Abadi D, Ahmad Y, Balazinska M, Cetintemel U, Cherniack M, Hwang J, Lindner W, Maskey A, Rasin A, Ryvkina E, Tatbul N, Xing Y, Zdonik S (2005) The design of the Borealis stream processing engine. Second biennial conference on innovative data systems research (CIDR 2005), ACM, Asilomar

Arasu A, Babcock B, Babu S, Cieslewicz J, Datar M, Ito K, Motwani R, Srivastava U, Widom J (2004) STREAM: the stanford data stream management system. Technical report, Stanford InfoLab. http://ilpubs.stanford.edu:8090/641/1/2004-20.pdf. Accessed 2 Jan 2015

Arasu A, Babu S, Widom J (2006) The CQL continuous query language: semantic foundations and query execution. VLDB J 15(2):121–142

Avgerinakis K, Briassouli A, Kompatsiaris I (2012) Smoke detection using temporal HOGHOF descriptors and energy colour statistics from video. Firesense Workshop, 8–9 Nov 2012, Antalya, Turkey

Barbieri D, Braga D, Ceri S, Della Valle E, Grossniklaus M (2010) Incremental reasoning on streams and rich background knowledge. The semantic web: research and applications, Lecture notes in computer science, Springer, vol 6088, pp 1–15

Barwise J (1981) Scenes and other situations. J Philosophy 78(7):369–397

Bradski GR (1998) Computer vision face tracking for use in a perceptual user interface. Intel Technology J 2(2):12–21

Blasch EP, Plano S (2003) Level 5: user refinement to aid the fusion process. In: Multisensor, multisource information fusion: architectures, algorithms, and applications. Proceedings of the SPIE, vol 5099, pp 288–297

Chandrasekaran S, Cooper O, Deshpande A, Franklin M, Hellerstein J, Hong W, Krishnamurthy S, Madden S, Raman V, Reiss F, Shah M (2003) TelegraphCQ: continuous dataflow processing for an uncertain world. Conference on innovative data systems research (CIDR 2003), ACM, New York

Della Valle E, Ceri S, Barbieri D, Braga D, Campi A (2008) A first step towards stream reasoning. Future internet symposium (FIS 2008), Vienna, Austria

Dey A (2001) Understanding and using context. J Ubiquitous Computing 5(1):4–7

Doorenbos R (1995) Production matching for large learning systems. PhD thesis, Carnegie Mellon University, Pittsburgh

Dougherty E, Laplante P (1995) Introduction to real-time imaging. Chapter what is real-time processing? Wiley-IEEE, New York, pp 1–9

Doulaverakis C, Konstantinou N, Knape T, Kompatsiaris I, Soldatos J (2011) An approach to intelligent information fusion in sensor saturated urban environments. European Intelligence and Security Informatics Conference (EISIC 2011), IEEE, Athens

Eid M, Liscano R, Saddik A (2007) A universal ontology for sensor networks data. In: IEEE international conference on computational intelligence for measurement systems and applications (CIMSA'07), pp 59–62

Endsley MR (2000) Theoretical underpinnings of situation awareness: a critical review. Situation awareness analysis and measurement. Lawrence Erlbaum, Mahawah

Ghanem T, Aref W, Elmagarmid A (2006) Exploiting predicate-window semantics over data streams. ACM SIGMOD Record 35(1):3–8

Hall D, Llinas J (2001) Handbook of multisensor data fusion. CRC, New York

Hall D, Llinas J (2009) Multisensor data fusion, book chapter, In: Handbook of multisensory data fusion: theory and practice, 2nd edn. CRC, New York

Horrocks I, Patel-Schneider P, Boley H, Tabet S, Grosof B, Dean M (2004) SWRL: a semantic web rule language combining OWL and RuleML. World Wide Web Consortium, Member Submission. http://www.w3.org/Submission/SWRL/. Accessed 30 Dec 2014

Kokar MM, Letkowski JJ, Dionne R, Matheus CJ (2008) Situation tracking: the concept and a scenario. Situation management workshop (SIMA'08), Military communications conference, IEEE, San Diego, pp 1–7

Kokar MM, Matheus CJ, Baclawski K (2009) Ontology-based situation awareness. Information Fusion 10(1):83–98 (Special Issue on High-level Information Fusion and Situation Awareness)

Konstantinou N, Solidakis E, Zafeiropoulos A, Stathopoulos P, Mitrou N (2010) A context-aware middleware for real-time semantic enrichment of distributed multimedia metadata. Multimedia Tools Applications 46(2–3):425–461

Lassila O, Khushraj D (2005) Contextualizing applications via semantic middleware. Second annual international conference on mobile and ubiquitous systems: networking and services (MOBIQUITOUS'05), IEEE Computer Society, Washington, DC, pp 183–191

Lawton G (2004) Machine-to-machine technology gears up for growth. IEEE Computer 37(9):12–15

Lefort L, Henson C, Taylor K (Eds) (2011) Incubator report. http://www.w3.org/2005/Incubator/ssn/wiki/Incubator_Report. Accessed 30 Dec 2014

Li D, Dimitrova N, Li M, Sethi I (2003) Multimedia content processing through cross-modal association. ACM International Conference on Multimedia, Berkeley

Li J, Maier D, Tufte K, Papadimos V, Tucker P (2005) Semantics and evaluation techniques for window aggregates in data streams. ACM SIGMOD International Conference on Management of Data (SIGMOD'05), pp 311–322

Llinas J, Waltz E (1990) Multisensory data fusion. Artech House, Norwood

Llinas J, Bowman C, Rogova G, Steinberg A, Waltz E, White F (2004) Revisiting the JDL data fusion model II. In Svensson P, Schubert J (eds). Seventh international conference on information fusion, pp 1218–1230

Neuhaus H, Compton M (2009) The semantic sensor network ontology: a generic language to describe sensor assets. AGILE international conference on geographic information science, Hannover

Papamarkos G, Poulovassilis A, Wood PT (2003) Event-condition-action rule languages for the semantic web. Workshop on semantic web and databases (SWDB 03), pp 309–327

Patroumpas K, Sellis T (2006) Window specification over data streams. International conference on semantics of a networked world: semantics of sequence and time dependent data (ICSNW'06). Springer, New York, pp 445–464

Perez J, Arenas M, Gutierrez C (2006) Semantics and complexity of SPARQL. International semantic web conference 2006 (ISWC 2006), pp 30–43

Polleres A (2007) From SPARQL to rules (and back). 16th International conference on world wide web (WWW '07), ACM, New York, pp 787–796

Schenk S (2007) A SPARQL semantics based on datalog. 30th annual German conference on advances in artificial intelligence, Wiley, New York, pp 60–174

Sheth A, Larson J (1990) Federated database systems for managing distributed, heterogeneous, and autonomous databases. ACM Computing Surveys 22(3):183–236

Sheth A, Henson C, Sahoo S (2008) Semantic sensor web. IEEE Internet Computing 12(4):78–83

Sundmaeker H, Guillemin P, Friess P, Woelffl S (2010) Vision and challenges for realising the Internet of things. European Commission, March 2010. doi:10.2759/26127

Toninelli A, Montanari R, Kagal L, Lassila O (2006) A semantic context-aware access control framework for secure collaborations in pervasive computing environments. The semantic web—ISWC 2006, Lecture notes in computer science. Springer, New York, pp 473–486

Trifan M, Ionescu B, Ionescu D, Prostean O, Prostean G (2008) An ontology based approach to intelligent data mining for environmental virtual warehouses of sensor data. IEEE conference on virtual environments, Human-computer interfaces and measurement systems (VECIMS 2008), pp 125–129

Wu Y, Chang E, Chang K, Smith J (2004). Optimal multimodal fusion for multimedia data analysis. ACM International Conference on Multimedia, New York

Chapter 6
Conclusions: Summary and Outlook

6.1 Introduction

This book aims at clarifying the LOD domain, from the point of view of LOD creation. We attempted to provide an overview of the aspects that altogether comprise the Linked Data generation problem. As we have analyzed throughout this book, the Web of Linked Data has come a long way, being nowadays at a mature stage. The LOD ecosystem is nowadays making its first steps into mainstream adoption. The domain presents numerous capabilities, and that there is huge potential for creating innovative applications.

The ultimate goal for tomorrow's Web applications is to embody a separate Linked Data Layer that can serve as the basis for intelligent processing. Figure 6.1 illustrates this vision, the materialization of which we tried to cover in this book.

The theoretical and technical challenges associated with this goal are introduced and analyzed in this book, followed by several examples and two complete use cases. Let us briefly recapitulate what we talked about in this book:

Chapter 1 set the tone for the entire book by introducing the vision of the Semantic Web and the Linked Data reality and discussing their prospects and solutions that they may bring to existing problems. The main Semantic Web concepts, key terms and prevalent issues that were investigated in this book were also introduced and defined.

Next, in Chap. 2, we moved from theory to practice by providing the technical background that materializes the concepts presented in Chap. 1. We presented the fundamental technologies that range from knowledge representation models that support inference to query languages and mapping formalisms. We also took a look at some of the most popular vocabularies that serve as a common language among Linked Data applications, therefore enhancing their semantic interoperability.

In Chap. 3 we provided a technical overview of the Linked Data designing and publishing processes. More specifically, we began with presenting how data can be

© Springer International Publishing Switzerland 2015
N. Konstantinou, D.-E. Spanos, *Materializing the Web of Linked Data*,
DOI 10.1007/978-3-319-16074-0_6

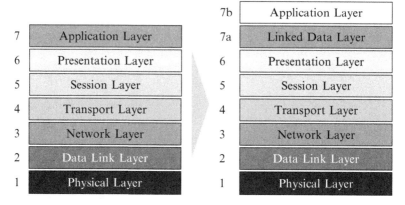

Fig. 6.1 OSI/OSI model and envisioned linked data interoperability layer (Decker 2010)

modeled and opened up to the public. We provided an overview of the most common technical solutions and widely used tools that can serve this purpose. Overall, in Chap. 3, we aimed to provide a first analysis of the sub-problems in which the LOD-publishing task is broken down, namely content modeling, opening, linking, and processing.

In Chap. 4 we analyzed how Linked Data can be generated from relational databases. We devoted this Chapter to the study of the more general problem of interfacing relational systems with Semantic Web applications, given the popularity of relational databases as storage solutions, often of huge data volumes that feed the vast majority of information systems worldwide. We gave several pointers to the related literature and we enumerated the main motivations that drive the relevant approaches as well as their benefits. We presented a categorization of approaches that map relational databases to the Semantic Web and highlighted the most popular working tools that extract RDF graphs from relational instances. We finally sketched a proof-of-concept use case scenario regarding how the data from an open access repository can be converted to Linked Data.

In Chap. 5, we analyzed the problem of processing and semantically enhancing data originating from sensor data streams, using the techniques and standards described in previous chapters. We introduced the basic concepts needed for our analysis, such as real-time processing, context-awareness, windowing and information fusion and we analyzed the challenges of setting up and maintaining a semantic sensor network. We finally presented a number of issues that have to be dealt with and demonstrate how an intelligent, semantically-enabled data layer can be materialized, leading to information that is effortlessly integrated from a set of distributed sensors.

Overall, this book aims at contributing in the Data Science domain, "the study of the generalizable extraction of knowledge from data" (Dhar 2013). The main contributions of this book include:

- A formal and informal introduction of the concepts dealt with in this book: semantics, ontologies, data and information, reasoners, knowledge bases, annotation, metadata, real-time, context-awareness, integration, interoperability, fusion, etc.

- A detailed survey of the state-of-the-art existing approaches and methodologies regarding technologies, tools and approaches in various aspects of the domain (see Chaps. 3 and 4).
- Complete discussions on the parameters for creating Linked Data from relational databases and sensor data streams (Chaps. 4 and 5).
- Two detailed architectural and behavioral description of two domain-specific scenarios for the Linked Data creation (see Sects. 4.5 and 5.6).

6.2 Discussion

The LOD ecosystem can be regarded as an open and distributed system. In these systems, a certain degree of *heterogeneity* is inevitable. According to (Bouquet et al. 2004), heterogeneity can be present in four levels: syntactic, terminological, conceptual, and semiotic/pragmatic. As discussed throughout this book, the Linked Data paradigm, offers a promising solution to reducing this heterogeneity, by defining relations across the heterogeneous sources, at all four levels of heterogeneity.

As shown in Chap. 1, the several disperse parallel efforts for LOD publishing have led so far to weaving a Web of Linked Data, as illustrated in Fig. 1.2. However, as far as it concerns how data can be put together in order to form a Linked Data repository, there are many steps and many components involved in processing at server-side the information generated in the system. There is no "standard" approach in configuring every component, and therefore, there can be *no deterministic manner* in setting the priorities in the resulting system's behavior.

Several things have to be taken into account in order to deploy successful LOD provision solutions, including, among others, *scalability*. This is an important system property in terms of (a) the volume of the produced, stored and processed information and (b) the number of the data sources. Information in systems involving several data streams can quickly rise to difficultly manageable volumes, especially when there are sensor/multimedia/social network data streams into play.

Additionally, in cases when *integration with third parties* is important (see Example in 4.5 regarding publishing information in the scholarly/cultural heritage domain), Semantic Web technology adoption is crucial in order to assure unambiguous definition of the information and the semantics it conveys.

Once the system's data repository is turned into a knowledge base, the possibilities for further exploitation are virtually endless, allowing the performance of e.g. analytical reasoning and intelligent analysis (such as data mining, pattern extraction) over the data. Far more possibilities are unleashed in the ways in which we can use the data, even in unprecedented ways. The numerous **benefits** that arise include the following (Konstantinou et al. 2013):

- *Increased discoverability*. Exposing the data as RDF graphs opens new possibilities in discovering it. Firstly, the data can be accompanied by a description over their contents (e.g. using the VoID vocabulary). Also, links can be realized towards instances in other parts of the LOD cloud, and also, inbound links are welcome as well, forming an interconnected island in the LOD cloud. Therefore, the dataset can be linked to other datasets, thus increasing its discoverability.

- *Schema modifications* require reduced effort. Creation of new relations is allowed without modifying the database schema. New classes and properties can be defined in the RDF schema that will apply to subsets of the repository data, without the need to modify neither the relational database schema nor its contents.

- *Synthesis* by means of integration, fusion, mashups. The end user/developer can perform searches spanning various repositories, from a single SPARQL endpoint. Also, it is possible to download parts or the whole data in order to combine it with other data and process it according to his/her needs.

- *Inference.* With reasoning support, new implicit facts can be inferred, based on the existing ones that are explicitly derived from the relational database. These new facts can then be added to the graph, thus augmenting the knowledge base.

- *Reusability.* Third parties can reuse the data in their systems, either by including the information in their datasets, or by providing reference to the published resources.

On the other hand, **technical difficulties** encountered in the course of things are expected, as they may be caused by a series of factors, as the following:

- *Multidisciplinarity.* As a multidisciplinary task, involving Computer Science and expertise in each specific domain, annotation alone is a task not to be underestimated, since it may require contributions from several scientific domains and close collaboration of the implementation team.

- *Technology barrier.* Tools are not as mature yet as to guide the inexperienced user, or warn for possible incorrect uses. For instance, when linking among datasets or performing a mapping (i.e. writing a R2RML file), the whole procedure may take place manually, in a plain text editor, without a way of assuring at design-time the validity of the result. This is, therefore, a step that requires the presence of an expert.

- *Result is prone to errors.* Even after producing the resulting graph, and in the case when it is syntactically correct, there is no validator service to perform an automatic check of whether the concepts and properties involved are used as intended. Errors or bad practices, even after being published can go unnoticed.

- *Concept mismatch.* It is not always possible to extract an RDF description from the stored values in the repository. Although RDF allows in general for more complex descriptions compared to the metadata model, identical mappings may not always be found and compromises have to be made.

- *Exceptions to the general rule.* When performing mass data curation or transformation tasks, it is expected that not all records can be simultaneously updated: Typically, changes made automatically will apply to the majority of the data, while the remaining portion will require manual intervention. Therefore, one can expect from automated procedures to cover the majority, but not the whole LOD generation process. It is expected that post-publishing manual interventions will be required.

6.3 Domain-Specific Benefits

Several domains are already benefitting from LOD adoption, paving the way for more to come, in the same, adjacent, or completely different domains:

- **Open Government Data**. Today, the open data movement is emerging across governments and organizations from all over the world. Organizations offer their data publicly, allowing its free reuse under minimal or no restrictions at all, fostering transparency, collaborative governance and innovation and enhancing the citizens' quality of life through the development of novel applications. The open data paradigm is embraced by several governments, which have set up web portals for the free distribution of data they own, such as the United States Government open data[1] and the United Kingdom Government public data[2] registries. However, the value of open data decreases if it is not released in open, globally understandable formats that are easily combined and linked with other open data. Therefore, efforts to integrate open governmental data to the LOD cloud are currently under way and are expected to intensify over the years to come.

- **Bibliographic Archives**. A huge wealth of human knowledge exists in digital libraries and open access repositories, typically accompanied by accurate structured metadata that allow cataloguing, indexing and effective searching on the aggregated content. This metadata is typically trapped inside monolithic systems that support Web-unfriendly protocols for data access. Linked Data is a technology that can bring a revolution in the way librarians catalog and organize items, allowing for direct reuse of the work of other librarians and transforming the item-centric cataloguing to entity-based descriptions. Several efforts at the national and regional level have already adopted the Linked Data paradigm for publishing their collections, notable examples including the Library of Congress[3] and the British National Bibliography.[4] Still though, there are steps that need to be taken for the seamless integration of bibliographic data in the LOD cloud and the interlinking of such datasets with LOD datasets from other domains. This interlinking is expected to give rise to novel applications that exploit library data in combination with other (e.g. geographical) data and ultimately, offer multiple benefits to libraries themselves.

- **The Internet of Things**. The billions of sensing devices that are currently deployed worldwide and the number of which is expected to continuously grow, already form a giant network of connected "things", the so-called Internet of Things (IoT). Every such device is uniquely identified and accessed through standard Internet protocols and fuel with data applications in domains as diverse as environmental monitoring, energy management, healthcare and home and city

---

[1] United States Government open data: data.gov

[2] United Kingdom Government Public Data: data.gov.uk

[3] Library of Congress Linked Data Service: id.loc.gov

[4] British National Bibliography: bnb.data.bl.uk

automation. The need for infusing intelligence in IoT architectures and applications calls for the use of ontologies and other Semantic Web technologies that support inference, i.e. the generation of new facts from already known ones. Such technologies can also facilitate the integration of already deployed IoT platforms and architectures that have been developed independently and nowadays constitute separate technological silos. The Linked Data paradigm fits nicely in the context of IoT systems, given that they can be easily integrated to the Web and their nodes are uniquely identifiable.

6.4   Open Research Challenges

According to (Dhar 2013) "Data science is the study of the generalizable extraction of knowledge from data". Provision of LOD is the first step, but it is not a goal in itself; it is rather a means to an end, the extraction of knowledge from the data.

Next, among the major challenges in the LOD ecosystem in general is the *consumption* of the available datasets in a manner that extracts intelligence and generates additional value. Under this term, we gather all actions related to *consuming*, as opposed to *producing*, which is covered in the hereby work. Overall, among the services that ease LOD consumption we can identify visualization, analytics, text mining, named entity recognition, etc. Dealing with Big Data, in cases when the datasets comprise millions or billions of facts, making the size of the dataset become a part of the problem itself, is an open research problem as well, with several approaches towards effective storage and querying. With the rise of several media channels, including social networks, blogs and multimedia sharing services, data is generated in growing rates and influences the decisions and actions of individuals in their professional and personal life. Therefore, the need for timely, accurate and efficient analysis of big data volumes is more evident today than ever before. The challenge of managing, querying and consuming large volumes of Linked Open Data has already begun to concern researchers and is expected to be a central issue in the following years.

Among the major challenges is the *quality assessment* of the Linked Data. The volume of being published on the Web as LOD is constantly increasing, however the quality of the datasets varies ranging from extensively curated datasets to crowdsourced and extracted data of relatively low quality. Therefore, among the challenges in consuming LOD is to determine the quality of the datasets, domain that attracts much attention lately (Zaveri et al. 2015; Kontokostas et al. 2014).

Big Linked Data is another research domain. The term *Big Data* has recently made its appearance, driven by the realization that novel techniques will be needed to handle the vast amounts of data generated in the near future. In general, there is not an agreed and formal definition of what exactly is (and what is not) big data, it is commonly characterized by different properties (all V's for some mysterious reason): volume, velocity, variety, value, veracity. Another common way to describe big data is when the data itself is part of the problem, being so large and complex that it becomes difficult to process it using conventional data processing applications. Of course, these problems arise in several domains, such as the Internet of Things, sensor networks, etc. As the authors in (Hitzler and Janowicz 2013) put it, it appears that Linked Data is part of the Big Data landscape, and go a bit further and claim that Linked Data is an ideal testbed for researching some key Big Data challenges. Ultimately, all V's have to be addressed in an interdisciplinary effort to substantially advance on the Big Data front (Hendler 2013).

This list, of course, is not exhaustive. Still, there are issues regarding e.g. privacy, legal aspects, integration and reconciliation from diverse data sources, to name just a few, leaving much room for research and making Linked Data one of the hottest topics in Computer Science nowadays.

References

Bouquet P, Ehrig M, Euzenat J, Franconi E, Hitzler P, Krötzsch M, Serafini L, Stamou G, Sure Y, Tessaris S (2004) Specification of a common framework for characterizing alignment. Technical report, Institut AIFB, Universität Karlsruhe, 2004. KnowledgeWeb deliverable d2.2.1v2

Decker S (2010) Linked data and the future Internet architecture: a motivation. Linked data in the future Internet at the future Internet assembly, Ghent, Belgium. http://fi-ghent.fi-week.eu/files/2010/12/1100-DeckerIntro.pdf. Accessed 7 Jan 2015

Dhar V (2013) Data science and prediction. Commun ACM 56(12):64. doi:10.1145/2500499

Hendler J (2013) Broad data: exploring the emerging web of data. Big Data 1(1):18–20

Hitzler P, Janowicz K (2013) Linked data, big data, and the 4th paradigm. Semantic Web 4(3):233–235

Konstantinou N, Houssos N, Manta A (2013) Exposing bibliographic information as linked open data using standards-based mappings: methodology and results. In: 3rd International conference on integrated information (IC-ININFO '13), Elsevier, Prague

Kontokostas K, Westphal P, Auer S, Hellmann S, Lehmann J, Cornelissen R, Zaveri A (2014) Test-driven evaluation of linked data quality. In: 23rd International conference on World Wide Web (WWW '14), ACM, New York, pp. 747–758

Zaveri A, Rula A, Maurino A, Pietrobon R, Lehmann J, Auer A (2015) Quality assessment methodologies for linked open data, Semantic Web journal. www.semantic-web-journal.net/system/files/swj414.pdf, under review. Accessed 30 Dec 2014

Printed in the United States
By Bookmasters